Environmental Footprints and Eco-design of Products and Processes

Series Editor

Subramanian Senthilkannan Muthu, Head of Sustainability - SgT Group and API, Hong Kong, Kowloon, Hong Kong

This series aims to broadly cover all the aspects related to environmental assessment of products, development of environmental and ecological indicators and eco-design of various products and processes. Below are the areas fall under the aims and scope of this series, but not limited to: Environmental Life Cycle Assessment; Social Life Cycle Assessment; Organizational and Product Carbon Footprints; Ecological, Energy and Water Footprints; Life cycle costing; Environmental and sustainable indicators; Environmental impact assessment methods and tools; Eco-design (sustainable design) aspects and tools; Biodegradation studies; Recycling; Solid waste management; Environmental and social audits; Green Purchasing and tools; Product environmental footprints; Environmental management standards and regulations; Eco-labels; Green Claims and green washing; Assessment of sustainability aspects.

More information about this series at http://www.springer.com/series/13340

Subramanian Senthilkannan Muthu
Editor

COVID-19

Environmental Sustainability and Sustainable Development Goals

 Springer

Editor
Subramanian Senthilkannan Muthu
Head of Sustainability
SgT and API
Kowloon, Hong Kong

ISSN 2345-7651   ISSN 2345-766X (electronic)
Environmental Footprints and Eco-design of Products and Processes
ISBN 978-981-16-3862-6   ISBN 978-981-16-3860-2   (eBook)
https://doi.org/10.1007/978-981-16-3860-2

This Springer imprint is published by the registered company Springer Nature Singapore Pte Ltd.
The registered company address is: 152 Beach Road, #21-01/04 Gateway East, Singapore 189721,
Singapore

Contents

About the Editor

Dr. Subramanian Senthilkannan Muthu currently works for SgT Group as Head of Sustainability, and is based out of Hong Kong. He earned his Ph.D. from The Hong Kong Polytechnic University, and is a renowned expert in the areas of Environmental Sustainability in Textiles & Clothing Supply Chain, Product Life Cycle Assessment (LCA), Ecological Footprint and Product Carbon Footprint Assessment (PCF) in various industrial sectors. He has five years of industrial experience in textile manufacturing, research and development and textile testing and over a decades of experience in life cycle assessment (LCA), carbon and ecological footprints assessment of various consumer products. He has published more than 100 research publications, written numerous book chapters and authored/edited over 100 books in the areas of Carbon Footprint, Recycling, Environmental Assessment and Environmental Sustainability.

Sustainability in Covid-19 Times: A Human Development Perspective

Carla Patricia Finatto, Camilla Gomes da Silva, Ana Regina de Aguiar Dutra, André Borchardt Deggau, Anelise Leal Vieira Cubas, Elisa Helena Siegel Moecke, Felipe Fernandez, and José Baltazar Salgueirinho Osório de Andrade Guerra

Abstract The COVID-19 pandemic can be divided into two waves: the first is associated with health problems, and the second with economic and environmental problems. However, it is necessary to analyze the existence of a third wave that, in the long run, can have a deeper impact on people's lives. This wave emerged from the virus capacity to accentuate social, economic, political, and cultural inequalities. In this sense, the novel coronavirus has profoundly affected efforts to achieve the 17 Sustainable Development Goals (SDGs), established in the 2030 Agenda, especially with regard to SDG 3, SDG 10, SDG 12 and SDG 16. In that spectrum, this chapter aims to demonstrate how the impacts of the COVID-19 pandemic affect compliance with the SDGs. The impact of this pandemic had been such that it is possible that it will mean the beginning of a new era, based on the need for global solidarity and the desire to pursue sustainable development paths. The COVID-19 pandemic provides an opportunity to propose new actions for a more sustainable world, drafting a recovery from economic and social crises that finds comprehensive solutions.

Keywords Social inequalities · economic and political inequalities · COVID-19 · sustainable development goals · just and inclusive societies

C. P. Finatto · C. G. da Silva · A. Borchardt Deggau
Research Centre On Energy Efficiency and Sustainability (GREENS), University of Southern Santa Catarina (UNISUL), Florianópolis 88015-110, SC, Brazil

A. R. de Aguiar Dutra
Center of Sustainable Development, University of Southern, Santa Catarina, Florianópolis 88015-110, SC, Brazil

A. L. V. Cubas · E. H. S. Moecke · F. Fernandez
University of Southern Santa Catarina (UNISUL), Florianópolis 88015-110, SC, Brazil

J. B. S. O. de Andrade Guerra (✉)
Centre for Sustainable Development (GREENS), University of Southern Santa Catarina, Florianópolis 88015-110, SC, Brazil
e-mail: baltazar.guerra@unisul.br

© The Author(s), under exclusive license to Springer Nature Singapore Pte Ltd. 2021
S. S. Muthu (ed.), *COVID-19*, Environmental Footprints and Eco-design
of Products and Processes, https://doi.org/10.1007/978-981-16-3860-2_1

1 Introduction

The Covid-19 pandemic has had profound and impacts on humanity over the past two years. The rapid transmission pattern with the wide geographical spread of the novel coronavirus (SARS-CoV-2) has seriously impacted different matrices of society in its three main pillars: economic, political and social. In addition to impacts on physical and mental health, societies are facing several environmental and economic challenges in areas ranging from quality of water, air, soil, biodiversity [1, 2] to public debt, quality of life, employability and waste management [3, 4].

The pandemic also pushes society to look to the past to analyze pandemics' economic impacts on socioeconomic inequality [5]. In this sense, the author highlights the Plague of Justinian (541–544) and Black Death (1347–1353) and presents evidence that demonstrates that previous pandemics have reduced inequality rather than increased it. However, he argues that in both cases, at least a quarter or even half of Europe's population may have perished—data on other continents is scarce. These conclusions, however, contradict the direction of most studies on inequalities linked to the COVID-19 pandemic, which points to a trend of increased inequality.

Reference [6] argue that pandemics have historically led to profound social transformations in societies, however, such transformations are not inevitable. The authors argue that previous pandemics led to reforms, pointing at examples such as the improvement in living and working conditions in Europe after the Black Death in the fourteenth century and improvements in British health systems after the cholera epidemic in 1832. However, such a system has not survived other epidemics of cholera in the following centuries.

The COVID-19 pandemic brings to light complex interconnected dilemmas of globalization, health equity, economic security, environmental justice, waste management, democracy and collective trauma, with a more significant impact the most vulnerable groups [7]. The pandemic has also highlighted existing infrastructure problems in areas such as healthcare, sanitation, housing, and access to essential items such as water, energy, and food. Such problems require government action to protect populations, especially in Latin America. Such processes increase the importance of understanding fundamental rights in decision-making [8].

In 2020, inequality has reached its highest level in the United States, with 1% of the population holding twice of the wealth of 90% of the population [9]. In this context, the literature points that, in contrast to what happened in previous pandemics, the COVID-19 crisis is increasing inequality instead of reducing it.

The COVID-19 pandemic provides an opportunity to propose new actions for a more sustainable world, drafting a recovery from economic and social crises that finds comprehensive solutions, using the Sustainable Development Goals (SDGs) as a framework—especially in order to avoid the stigmatization of already marginalized groups [10, 11].

In this objective, it is worth highlighting SDG3, set by the United Nations in 2015. SDG 3 establishes the health and well-being of humanity as its main focus and priority. To achieve this goal, 13 targets were set to measure progress.

SDG 3 stands out among the goals by directly connecting with groups of people who are vulnerable to the impacts of the COVID-19 pandemic, mainly in targets 3.8 and 3.d, which refer to the search for the well-being and health of all human beings. According to Ref. [12], the current pandemic overloads health systems and has had several adverse factors impacting the various goals of SDG 3 differently. Besides, the first four goals of SDG 10 should be emphasized, as they advocate the reduction of inequalities in the world, which are currently being intensified.

In short, all of these goals propose universal health coverage; early preparation for possible global health risks; increase in the income of the poor population; social, economic and political inclusion; promoting standards that reduce inequality; and the adoption of policies to achieve greater equality among all.

In the environmental sphere, changes in behavior and consumption are necessary to achieve sustainability, including less resource consumption and guaranteeing spaces for future generations [13]. On the COVID-19 pandemic, residential and hospital waste production has increased, hindering efforts to comply with SDG 12 by creating new points of pollution on air and sea [14].

There is an opportunity for that transition in a post-COVID-19 world. The Ref. [15] has called for solidarity, not stigmatization. However, the organization has submitted no substantive guidance on how countries can take public health measures that achieve health protection while respecting human rights. This is mainly due to two factors. First, there is still not enough information on the best way to contain this pandemic. Second, factors such as sex, race, class, disabilities, ethnicity, and other axes of identity are still relevant to determine inclusion in society and, by extension, vulnerability to a pandemic [8].

Therefore, the COVID-19 pandemic and measures taken to fight it can have serious long-term consequences that would affect compliance with the SDGs [16]. McNeely [17] adds that sustained economic growth and the globalization of human movement, interconnections, finances, trade, and economic investments are linked with the completion of the SDGs by 2030. In the face of the COVID-19 pandemic, sustainable actions are being severely limited, failing to include all and affecting specifically isolated populations.

In that context, joint action between all social actors is necessary to achieve a fair and equalitarian society. The SDGs provide a fundamental framework in which jobs, social equality and economic concerns will be addressed in the coming recovery [18]. In this context, the structural violence existing in Brazil and in the world becomes increasingly evident, which, according to Elavarasan and Pugazhendhi [19], is an avoidable deficiency of fundamental human needs, which is based on a conjuncture of extreme social inequalities, where part the population is excluded and does not have access to rights, which makes room for the devaluation of life and the trivialization of death and impunity.

In this perspective, Ref. [20] argues that diseases, in general, are not democratic, as their incidences are determined by income, housing, age, gender, and race. In the case of this pandemic, that is no different due to the vulnerable populations that are already in the risk group.

This chapter aims to demonstrate how the impacts of the COVID-19 pandemic imply compliance with the SDGs, represented in Fig. 1. In addition, we prove that the response to such an emergency must be a constant construction in order to make society more egalitarian, primarily through cooperation resources, technologies, transparency in the dissemination of data, the responsibility of decision-makers, entrepreneurs and civil society, high investments in health services and political-economic intentions of governments [21].

The following sections will address the impacts of Covid-19 on the global health of the world population, with repercussions on health (SDG 3), social inequalities (SDG 10), the promotion of just, peaceful and inclusive societies (SDG 16) and, finally, the influence of the generation of waste by the novel coronavirus on responsible consumption and production and the balance of the environment (SDG 13).

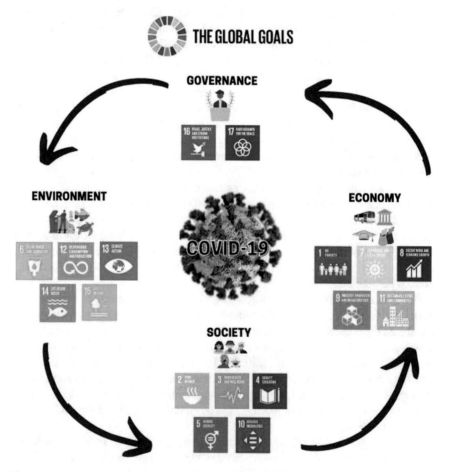

Fig. 1 Representation of the purpose of the chapter (*Source* Own elaboration)

2 Methodology

The integrative review method employed in this research consists in synthesizing knowledge from primary findings, obtained from indexed databases, such as Scopus, ScienceDirect, Emerald, Web of Science and Google Scholar, representing underground literature. The steps developed to unfold this chapter are represented in Fig. 2.

The period used as a research filter in the databases was 2017–2021 and the terms used for this research were: "global health" AND Covid-19; "SDG 3" AND Covid-19; Sexual AND "reproductive health" AND "family planning" AND "Covid"; "tobacco control" AND Covid 19; Covid 19 AND "social inequality" AND "economic inequality" AND political inequality; "Covid 19" AND "social inequality" AND "Economic inequality" AND "political inequality"; Covid 19 AND "social inequality" AND "economic inequality" AND "political inequality"; Covid-19 AND people AND justice; Covid-19 AND consumption AND Waste AND production AND environment ".

In these first searches in the databases, we found 2,820 primary findings and the used sample resulted in 133 scientific articles, distributed as follows: Scopus (39%); Sciencedirect (23%); Web of Science (11.5%); Emerald Insight (6.5%); Google Scholar (20%), in 56 indexed journals, the most cited journals were:: Environmental Research; Science of the Total Environment; International Journal of Sociology and Social Policy; The Lancet; Plos Medicine; The American journal of tropical medicine and hygiene; Sustainability; Journal of Pharmaceutical Policy and Practice; BMJ global health; Economies; Journal of global health; Public Health Nursing; Journal of Hospice & Palliative Nursing; Social Sciences; Research in Social and Administrative Pharmacy.

In the following section we will explore four important topics for this chapter: The impacts of Covid-19 on the global health of the world population, with repercussions

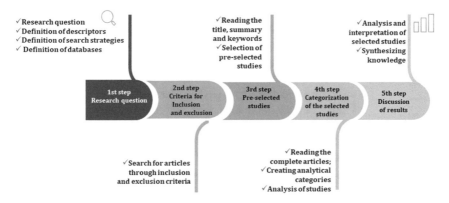

Fig. 2 The methodological steps (*Source* Own elaboration)

on SDG 3; Covid-19 and its impacts on the social, economic and political inequalities of the world populations; the spread of COVID-19 around the world and the promotion of fair, peaceful and inclusive societies; the influence of waste generation by Covid-19 on responsible consumption and production and on the balance of the environment.

3 Results

3.1 The Impacts of Covid-19 on the Global Health of the World Population, with Repercussions on SDG 3

The Covid-19 pandemic has had profound impacts on humankind over the past two years. The emergence of the COVID-19 has tested public health systems globally, impacted the core of neoliberal ideology, alerted to the issue of climate change with the temporary interruption of human activities, and caused health systems to be rethought in several countries. Reference [22] highlights that washing hands, practicing social distance and staying at home are the preventive measures implemented to contain the spread of COVID-19. However, although easy to follow, these measures highlight the tip of a socioeconomic iceberg and a socio-technological imbalance. Reference [23] argues that unsustainability had facilitated the spread of the SARS-Cov-2 in aspects such as harmful interactions between ecologic and socioeconomic systems and human invasions of natural habitats.

Reference [24] predicted that the overload of health systems caused by the pandemic would have the potential to cause up to 1.2 million children's deaths and 600.000 mothers' deaths that impacts the SDG 3 and its targets 3.1 (referring to infant mortality) and 3.2 (maternal mortality). Those scenarios might change depending on how long this pandemic scenario lasts, which is still unclear at the time of writing.

According to Ref. [25], the COVID-19 pandemic can be characterized as Syndemic since it affects different geopolitical contexts in a specific fashion and does not act alone, but it is compatible with other diseases [26]. The strategies to fight the COVID-19 pandemic have varied dramatically among different nations, and some developing countries in Sub-Saharan Africa have dealt with it much better than governments of developed countries, such as the United States.

In the case of target 3.3, attention should be paid to the health of marginalized and stigmatized populations, such as LGBTQI + and female workers, who end up encountering additional difficulties. Reference [27] highlight that these groups are often outside the State's social protection and were sometimes unprotected even before the COVID-19 pandemic. This is the case of some countries in Africa where prejudice was already a social problem even before the pandemic.

In turn, the research by Ref. [28] punctuates a wide range of impacts from the COVID-19 pandemic, including impacts on mental health and the treatment of non-communicable diseases, including the agenda for target 3.4. With regards to mental health, Clay and Parker [29] point out that periods of social isolation can have several impacts on health, leading to an increase in alcohol misuse, relapse and, potentially, the development of alcohol use disorder, putting even more pressure on addiction and drug and alcohol services, and health services in general, during and after the pandemic, a concern highlighted in target 3.5. Reference [30] point out that the increase in alcohol use was widely documented during other crises, particularly among people with anxiety and depression. Target 3.6, which deals with deaths and injuries from road accidents, Refs. [31] argue that social isolation has led to reducing minor traffic accidents, however, fatalities are still significant.

Target 3.7, which aims to ensure universal access to sexual and reproductive health services, including family planning, can be overwhelming and local and national measures imposed by countries in an attempt to control the spread of Covid-19 have also affected sexual and reproductive health services. Reference [32] argues that such changes will increase the needs for modern contraceptives, unwanted pregnancies, unsafe abortions, maternal and neonatal deaths and might incentive other harmful practices, such as female genital mutilation and child marriages in developing countries.

The need for contraception and reproductive planning during the COVID-19 pandemic is fundamental. Reference [33] claims that sexual and reproductive health (SRH), especially contraception, is an essential service. Besides, women, health professionals, legislators and society should be encouraged to consider SRH services as a priority.

Reference [34] argues that family planning has become even more important in light of the pandemic since COVID-19 has impacts on individuals and couples' rights. Furthermore, family planning is closely associated with fundamental human rights, such as equity, equality and universality. There is a need for continued national commitment and investments in family planning, especially during these difficult times when the vital economy is compromised due to the global crisis caused by the pandemic.

Still in relation to target 3.7, it is worth noting that sexual and reproductive health (SRH) pose a significant public health challenge problem during epidemics. A recent survey conducted in Tunisia revealed that up to 50% of SRH clinics in the country have been reduced or suspended activities since the emergence of COVID-19 [35]. Reference [36] points out the impacts of the pandemic on the health system with interruptions in the regular provision of SRH services, such as prenatal and postnatal examinations, safe abortions, contraception, HIV/AIDS and sexually transmitted infections, highlighting the need urgency of the scientific community to generate solid clinical, epidemiological and psychosocial behavioral links between COVID-19 and SRH [37]. In this pandemic scenario, Ref. [38] recommend that countries include family planning and reproductive health services in the essential service package and develop strategies to ensure that women and couples can exercise their reproductive rights COVID-19 crisis.

Another possible point of concern that impacts directly target 3.7 is the possible impacts of SARS-CoV-2 on fertility. Reference [39] explain that while, based on the current evidence, the probability of transmission of SARS-CoV-2 through the seminal fluid is very low, the data are still very limited to be sure of the impacts of SARS-CoV-2 infection on male reproductive hormones and in the semen parameters.

Social isolation measures designed to slow the spread of COVID-19 and reduce risk to medical staff also limit customers' involvement with SRH services. Several international agencies have estimated that the pandemic's long-term consequences will include a negative impact on the sexual and reproductive health needs of women and girls in vulnerable communities [40]. After experimental studies, Ref. [41], demonstrated that the pandemic had caused an increasing desire to postpone or prevent pregnancy while creating barriers to contraceptive services. This can lead to an increase in unwanted pregnancies, particularly among people who had difficulty buying food, transportation and/or housing during the pandemic.

Other studies reinforced the issue of universal health coverage, such as those developed by Ref. [12], on the role of pharmaceutical assistance in the midst of the COVID-19 pandemic. References [42] and [43] addressed health spending, and Ref. [44] focused on the study of a universal social protection system.

Reinforcing this literature, Ref. [45] studied the responsiveness of health systems, presenting four main propositions: integration, financing, resilience and equity. These proposals took into account the non-alignment between the rates of preparedness to respond to pandemics that predicted great response capacities for countries such as the United States, which ended up having markedly poor crisis management compared to other countries. Although the United States has an impressive range of public and private laboratories, innovative pharmaceutical and technology companies, high-capacity institutes in the field of public health, among other facilities, the country has a highly fragmented public health system. Furthermore, Ref. [46] propose the need to create a global fund with the objective of strengthening health systems, so that they are able to face new global pandemics, at the same time that such a fund would contribute to achieving the SDG 3.

Target 3.9 addresses two important points: deaths and diseases due to the use of dangerous chemicals, contamination and pollution of air, water and soil. Rume and Islam [47] indicate that in the current context, there has been an improvement in air quality in different cities around the world, a reduction in greenhouse gas (GHG) emissions, and a reduction in water and noise pollution, helping to balance the ecological system. However, Dharmaraj et al. [48] point out that despite the positive aspects, it is important to highlight the negative consequences of COVID-19, such as increased hospital waste (effluents, masks, and gloves).

In addition to the discussions made in the previous paragraphs about the repercussions of Covid-19 on global health and SDG 3, regarding targets 3.1 to 3.9, we will now dedicate ourselves to establishing relationships with the complementary goals, which are 3.a, 3.b, 3.c, 3.d. With regard to smoking, studies have shown the correlation between the prevalence of tobacco use in adults and mortality from COVID-19 worldwide and, according to Ref. [49], factors such as low immunity and a history of respiratory conditions lead to decreased lung capacity, which makes smokers a risk group for the novel coronavirus [50]. Diseases caused by the use of tobacco, such as cancer and cardiovascular diseases, act as comorbidities, aggravating symptoms in COVID-19 infection. Reference [51] highlight that nicotine exposure is linked to cardiopulmonary vulnerability to COVID-19 and tobacco use may be a potential risk factor for contracting viral infection and manifesting more serious symptoms. Another aggravating factor is the increase in infections due to the sharing of smoke and the release of droplets of steam and smoke, putting the community at risk [49].

The correlation between the prevalence of smoking in adult men and mortality from COVID-19 is higher in low-middle income countries. Reference [52] also highlight the importance of strengthening smoking control policies to reduce the impact of the COVID-19 pandemic in these locations. Reference [53] underlines that with public health priorities aimed at controlling the pandemic, there is a concern that tobacco control will be set aside, despite the fact that it causes millions of deaths per year. Reference [51] also suggest taking advantage of this pandemic moment to break the nicotine dependence cycle and accelerate national tobacco control programs to achieve a tobacco-free world. Tobacco control is a greater challenge than ever in the context of the pandemic COVID-19. Reference [54] emphasize that decision-makers must be vigilant to ensure that public health practices are consistent and compatible with the principles of the World Health Organization (WHO) Framework Convention and the SDGs for Tobacco Control.

There is also interest in creating teams that will help create a better understanding of disasters and health-related risks, aiming to strengthen local decision-making. Authors such as [55] show how these teams could help create a deeper understanding of the behavior of COVID-19 [56]. Osingada and Porta [57] also emphasize the importance of proper training of health professionals so that their formation will include a holistic approach, addressing concerns expressed in the SDGs. Reference [58] address gender equity to strengthen the workforce and the role of nursing in facing the crisis caused by COVID-19 [59].

3.2 COVID-19 and Its Impacts on the Social, Economic and Political Inequalities

In addition to all of its impacts on health, the COVID-19 pandemic has also significantly impacted social, economic, and political issues worldwide, especially in developing countries. The global economy has slowed as most companies have been affected. This situation is made worse by some countries' policies in response to this pandemic, from social distancing to blockade policies [60].

This social shock is increased when companies are forced to reduce their production to save their costs [61]. In this sense, Ref. [62] believe that poverty and food security can grow as the Covid-19 outbreak progresses. Reference [63] state that the virus has increased pre-existing inequality levels and has hit socially vulnerable people harder. In the same vein, Ref. [64] highlight that billions of lives worldwide were directly or indirectly impacted by the pandemic COVID-19, revealing and aggravating the social and economic inequalities that have emerged in recent decades.

This situation has expanded existing divisions by income, age, gender and ethnicity, exacerbating many existing inequalities and opening up new fissures, especially among those whose jobs cannot be done from home, which are often correlated with existing inequalities (for example, by income). Younger and low-income workers are much more likely to lose their jobs and suffer a reduction in earnings during a lockdown. Self-employed workers and workers with less secure employment contracts are also more likely to report negative impacts. Key workers, who generally face more health risks, are more likely to receive lower wages, be women, and belong to some minority ethnic groups.

In general, health impacts have been uneven, with higher mortality rates among certain occupations, ethnic minority groups and more inferior locations. Children in more impoverished families were more deeply affected as schools closed, and those who would have joined the labor market in 2020 face the potential for long-term negative implications due to the collapse of the labor market. In contrast, individuals with a higher level of education and higher income are more likely to work from home, homeschool their children and have savings to cover unforeseen expenses [65].

Governments must play a central role in leading all response preparations, coordinating efforts to avoid creating a vacuum to be filled by competing political parties. Most importantly, decision-makers need to address social inequalities and provide social protection and health systems for all, especially for disadvantaged populations, to mitigate economic vulnerabilities [66]. For [67], the COVID-19 pandemic intensified the economic and social problems that society had faced for decades, but the crisis also presents a unique circumstance of social cooperation opportunities.

In countries where trust in the public sector and the state was already weakened, the spread of misinformation related to a public health problem becomes more prominent and faster with the global proliferation of social media. That behavior was observed during the COVID-19 pandemic, accompanied by an unprecedented wave of disinformation described as damaging as the pandemic itself. Misinformation is understood as false information that is disseminated regardless of its intention to deceive [68].

Reference [66] also warn about false statements made by politicians, high-level elected officials, celebrities, prominent public persons and the general public about the spread of diseases and medicines, such as the ideas that saunas, hairdryers and exposure to the sun could prevent contamination from COVID- 19.

At this point, it is vital to highlight the relationship between COVID-10 and SDG 16, which, for [69] occurs mainly in the face of political polarization, which is a cultural barrier to coordinated action within countries. The polarization between citizens occurs in two forms: attitude polarization, which concerns supporters who take extremely opposite positions,and affective polarization refers to supporters who do not like and distrust those who present extremely opposite opinions. Affective polarization has political consequences such as decreased confidence, the privilege of party labels over political information, and the belief in false information, which can undermine social and economic relations and harm public health. Reference

[70] also claim that political polarization can be exacerbated by individuals' different news sources with different political inclinations.

Reference [71] show evidence that COVID-19 impacts different populations in widely varying ways: the poor, elderly, black and indigenous populations, as well as those who live with comorbidities, tend to fall ill and die at the highest rates. The social detachment guidelines changed millions of people to work from home and millions more lost their jobs, even when domestic workers, predominantly women, blacks, indigenous people and people of color, were asked to put the lives of their loved ones on the front lines.

In the United States, these biological, social and economic crises were punctuated by civil unrest, as millions took to the streets for racial justice, observing the unequal impacts of the pandemic. [72] state that the District of Columbia's food insecurity reveals a history of unequal access to food that was only amplified by the vulnerability of food supply chains during the COVID-19 pandemic. New opportunities for food access are being presented by advances in urban agriculture and other innovations in food production. These techniques could offer urban communities sustainable alternatives to access to food that simultaneously meet local food security and green infrastructure needs. But they also bring persistent socio-political barriers into greater focus.

The COVID-19 pandemic and the policies of social isolation it entails have exacerbated these barriers, making conventional solutions for access to food inadequate to meet its well-intentioned objectives. The ability to order groceries and household products on mobile devices, for example, is still unknown for a large part of the population. The profound disadvantages of marginalized populations and the isolating nature of structural racism. Contrary to the market-centered focus of traditional food access policies, such as public–private partnerships, disparities in access to food and resulting inequalities in food security are persistent problems in cities in the United States [73], [72].

With the arrival of COVID-19 in Brazil, a crisis scenario that incorporated economic, social and political aspects became quite visible. This scenario generated unemployment, poverty, and hunger and exposed several vulnerabilities that were worsening in recent years before the pandemic, making it easy for COVID-19 to find fertile ground in Brazil for its dissemination and community transmission. The impacts of the suspension of many commercial activities and other economic sectors were quickly felt socially and economically.

Some of the actions carried out by the Brazilian government included the payment of emergency aid (US$120/month for five months) and exemption from the payment of the energy bill for vulnerable people, the release of funds for programs of direct purchase of food from family farming, delivery of school feeding kits directly to students despite the closing of schools and publication of sanitary rules for the operation of restaurants [74]. The study carried out by UNICEF and the Brazilian Institute of Public Opinion and Statistics (IBOPE) from July 3 to 18, 2020 showed that during the pandemic, one in every five Brazilians aged 18 or over (33 million) experienced an episode of not having money to buy food when their income was over. This study also reports that about nine million Brazilians were unable to have a meal because there was no food or money to buy it [75].

In Indonesia, a large part of the population was unable to survive and meet their basic needs in the larger cities and returned to the small villages where they used to live, creating new problems such as the threat of virus transmission in addition to social and economic problems. With unemployment, they suffered an additional burden: the lack of natural resources, competing with local populations to use natural resources to satisfy their life needs [61].

In Greece, economic problems have greatly influenced the structure and resources of the country's health system. In addition to economic challenges, the country faces a refugee crisis, characterized by many critical points of overcrowding and tensions with neighboring Turkey. From an economic point of view, the impact of the COVID-19 outbreak can be worrying. As tourism is one of the country's main industries, prolonged travel restrictions during the summer can significantly affect the economy. On the other hand, the praise that Greece receives at the international level for the country's response to the outbreak and for the protection of public health is expected to preserve its reputation and attract tourists as soon as measures are lifted. Besides, the government-financed small and medium-sized enterprises that were affected by the pandemic and subsidized dismissed workers [76].

South Korea's reaction to COVID-19 represents a positive alternative to the dominant form of oligarchic government that prevails in Euro-American societies. The ruling elite implanted state power in ways that used this environment to continue previous patterns of domination that continually expanded surveillance, extending vital data extraction techniques for commercial and political purposes [77]

The state of Bangladesh has proved unable to implement policies of isolation or a fair and effective aid program. As a test of the state of Bangladesh, the pandemic served to highlight not only its institutional weaknesses but the contingency of citizens' compliance with policies seen as unfair and unfeasible; in such a context, the state could have acted more coercively but preferred to be tolerant before discreetly abandoning such policies. Both non-compliance and indulgence only make sense in light of the power of moral economy constructions of the state's role in subsistence crises. A vital challenge remains the ability to control political clientelism in the public interest [78].

According to [79], the debate about the pandemic highlighted the logic of the discourse that guided the various voices in Italy. There are two main perspectives guiding the public debate: the biomedical and the economic. The first defended biological life as the ultimate element of truth and legitimacy of government action. The second view is based on the justification of a careful cost–benefit calculation and the protection of the interests of the *homo oeconomicus*. However, the debate lacked a social perspective capable of placing dignity and human rights as a compass of intervention. Behind an apparent impartial universalism that would boost the biomedical and economic logic, there is a form of discrimination and lack of protection for specific sectors of society, particularly the marginal ones.

In Latin America, Ref. [80] criticize neoliberal economic postulates and the policy of fiscal austerity, which according to these authors provides privileges to elites at the expense of immense social damage that would have been exacerbated and accentuated during COVID-19. In Chile, Ref. [81] confirm the hypothesis that regional inequalities within countries impact the effects of the COVID-19 crisis. Furthermore, they reinforce other studies that point to the socioeconomic issue as a crucial problem exacerbated by the pandemic's challenge.

In India, Ref. [82] affirm that the health system can be harmed if there is an excessive hospitalization, due to the lack of adequate infrastructure and specialist doctors in relation to the high number of potential patients in need of intensive care, given the already low expenditure with the public system healthcare, which is 1.28% of total government revenue. In addition, the catastrophic cost of testing and treatment for patients who are not entitled to government insurance and subsidies will further increase debt and poverty in the country. The COVID-19 pandemic is likely to cause an economic crisis in India, as around 4% of GDP is expected to be lost during the management and recovery phase.

According to Dutta and Fischer (2020), as a low-income country, India also relies moderately on aid and funding from international organizations to control the spread of the disease. And the continued loss of jobs and the influx of migrant workers after the blockade phase reflected the government's lack of sustainability of civilian employment. This shows that an adequate emergency and preparedness plan is essential to avoid catastrophic losses in the financial sector and in the already needy health sector, which India must integrate into its basic public health program.

In the European Union (EU), the outbreak of the COVID-19 pandemic put intense pressure on providing a timely and coordinated response, capable of containing the disastrous economic and social effects of the pandemic in EU member states. In this situation, supranational institutions and their models of action were under pressure, seeming unable to make a decision for the ongoing crisis [83]. In the review carried out by [84], it is stated that the EU has implemented numerous strategies to answer emerging questions. Member States have taken measures such as closing borders and significantly limiting people's mobility to mitigate the virus's spread. An unprecedented effort to coordinate crises between Member States has facilitated the purchase of equipment and other medical supplies. Attention has also been focused on providing substantial research money to find a vaccine and promote effective treatment therapies. Financial support was made available to protect workers' wages and businesses to help facilitate a return to a functional economy.

In Palestine despite the serious social, health, political and economic impacts of the COVID-19 outbreak on Palestinians, Ref. [85] claim that a positive aspect of this pandemic is that it has revealed the dangers and shortcomings of traditional education centered on in the teacher who colonizes the students' minds, compromises their analytical skills and, paradoxically, puts them in a system of oppression that audits their ideas, limits their freedoms and restricts their creativity. Although the Israeli occupation proved to be an obstacle in the face of the Palestinian government's attempt to combat and contain the COVID-19 crisis, online education, the only arena that escapes this colonial system, has forced many instructors to give up their grip on the education process and to create a more collaborative educational environment that is based on dialogue, research and flexibility of curriculum content.

Although the number of COVID-19 cases in Africa is relatively limited for the time being, the pandemic and restrictive measures to reduce the virus can have important implications for the level of human security. They can cause economic decline and increased poverty, authoritarianism, urban violence and increased social inequalities [86].

According to [87], in Africa, the COVID-19 pandemic is responsible for a health crisis in frontline health professionals' victimization and the growing number of cases and deaths. At the same time, it causes a social crisis with the violation of human rights, the murder of citizens by the security forces and an increase in crime. This, in turn, exacerbates social inequalities, the breakdown of families, cases of social unrest and general impoverishment. An economic crisis also emerges, manifested by a decline in GDP and mass unemployment. A political crisis is demonstrated on the implementation of measures that may not be appropriate for Africa.

Concerning the issue of inequalities, [88] maintain the need to empower the poorest to face the richest lobby. For those authors, a series of dismantling social policies during the pandemic highlighted the fragility of the poorest, and the authors believe that the discipline of law can offer means to respond to these inequalities. Also, it should be noted that, as argued by the authors mentioned earlier in this study, not all groups are affected equally by the pandemic. In this sense, it is crucial to observe the impacts of COVID-19 on each of the most vulnerable socioeconomic groups.

The database research also presented chapters with contributions on these issues. Reference [89] discussed especially the effect of COVID-19 on undocumented immigrants. This group is especially vulnerable in some countries because they are reluctant and fearful when seeking specialized health services as they are in an illegal condition. From a social and political standpoint, the issue of gender is also affected by the COVID-19 pandemic. Reference [90] carried out a case study in Gambia on the impacts of the pandemic in the field of women's education, concluding that restrictions on the functioning of educational institutions had a disproportionate weight on the vulnerability of women in that society, which the authors portrayed as traditionalist and patriarchal.

Staying in a socio-political perspective, Ref. [91] consider that in addition to all the issues raised by the virus, misinformation, distrust and denialism emerged as a social situation in the same way as they did in the AIDS epidemic of the 1980s. Reference [68] corroborate this concern through a case study referring to Lebanon.

In Canada, Ref. [92] observed the existence of patterns of discriminatory behavior related to COVID-19 and noted that non-whites, younger people and health workers were more likely to face this type of behavior. However, on the other hand, less than a fifth of the study participants reported this type of behavior. Reference [93], in turn, addresses concerns about the situation of children and adolescents by stating "that they have had their current lives and their imaginary futures changed beyond recognition as a result of the virus." According to the author, more than 1.5 billion children and young people, or 87% of the world's student population, have stayed away from universities and schools. However, this group would be much more vulnerable to austerity policies. Young workers with precarious jobs also did suffer disproportionately from the effects of the interruption of economic activity.

3.3 The Spread of COVID-19 Around the World and the Promotion of just, Peaceful and Inclusive Societies

As we have previously states, the crisis generated by COVID-19 has accentuated the immense inequalities that exist in the world, which has affected and affects more intensely the historically most vulnerable and under-valued social groups, such as the elderly, women, indigenous peoples, homeless people and people living in impoverished areas or without access to the conditions necessary to face the disease, as well as small and medium-sized companies and the informal sector [94], [10].

In short, the novel coronavirus has damaged lives and livelihoods around the world. The impact of the pandemic on human lives is evident, but the effects on the global economy and sustainable development are also a concern [95]. In the first three months of 2020 alone, the [11] revealed that 5–25 million jobs were lost.

With the multipolarity of the globalized world, where different cultures, policies and socioeconomic realities coexist, the skills to deal with the present crisis are still lacking [96], which makes it difficult to contain the disease globally. In underdeveloped countries the impact is even more intense. That is why simple actions such as hand washing with soap and water, together with social distancing—measures widely advocated as a way to prevent the spread of COVID-19—become almost impossible for a considerable portion of the population, especially for the homeless or people living in impoverished locations, who lack basic sanitation [97]. The number of people belonging to vulnerable groups is increasing as job losses increase [98].

Thus, the COVID-19 pandemic increases the structural violence that exists in the world, because some people will have its right to protection from the virus guaranteed and others not, directly affecting SDG 11 (sustainable cities and communities) and SDG 10 (social inequalities in and between countries) [99].

In the world taken hostage by COVID-19, therefore, the importance of a systemic logic to solve the sustainability challenges arises. The transmission of the virus from animals to humans occurred through environmental degradation [100], while the spread of the virus among humans is closely linked to inequality: people living in poverty and those with underlying health problems - which are correlated - are the most vulnerable [101].

In addition, [13] argue that the virus is even more potent in the age group of 60 years or more, in countries where the population is already exposed to pollution and in countries that host the majority of international travelers in a global perspective. This is because the pandemic is no longer local but worldwide and requires more planned global cooperation, big data technologies and networks for decision making, transparency for data dissemination, responsibility of decision makers, businessmen and civil society, high investments in health services and economic intentions of governments, with the aim of lowering the cost of health and recovering the post-pandemic world economy of COVID-19 [10].

Regarding refugees, the pandemic has also made migrant workers more vulnerable to discrimination and xenophobia. In the same sense, Ref. [100] point out that overcrowding in the fields, settlements and shelters of this group is more prone to crowding people—a factor that considerably increases the level of infection.

Like any other disease, the first effects are felt in health systems, threatening SDG 3 (Good health and well-being). Hospitals and other health facilities in many countries are overburdened, leading to a lack of beds for medical care. The lack of equipment and infrastructure in weak health systems implies high mortality rates, especially in emerging economies, as in Brazil [102]. In this context, SDG 9 (Building resilient infrastructures, promoting inclusive and sustainable industrialization and fostering innovation) becomes an effective guide to safeguarding public and private institutions in preventing crises and systems collapse.

From another perspective, in the field of sciences there are unprecedented levels of collaboration in health research, a fact that touches SDG 3. Medicines and vaccines have never been developed so fast—giving a clear indication of human capacities to manage and develop innovative solutions in times of pressure with agility and international cooperation. In this way, increased North–South and South-South cooperation at various levels, together with a global technology facilitation and coordination mechanism to contain and find a cure for COVID-19—benefiting SDG 17—will provide learning for the years to come [97].

It is undeniable that the pandemic has reinforced the connections between health, environment and economy in developed and developing countries alike, in the same way that the SDGs apply for all countries [103]. Therefore, the response to the pandemic cannot be separated from the SDGs. In fact, achieving these goals will put us on a firm path to address global health risks and make us better prepared to face new emerging infectious diseases [104]

Still, it cannot be neglected that COVID-19 increases the likelihood of conflict (both within and outside borders) and, therefore, undermines the goal of global peace and justice (SDG 16: Peace, justice and strong institutions). The pandemic highlights the links between SDGs, especially between drinking water and health. in a scenario where a large percentage of the global population does not have access to adequate sanitation (SDG 11) and drinking water (SDG 6), and still face situations of poverty (SDG 1), hunger (SDG2) and inequalities (SDG10).

SDG 4 (Quality Education for All) was also affected, since COVID-19 demanded the closure of schools in order to prevent the spread of the virus, denying access to education, especially people in rural areas and populations in developing countries, which do not have access to basic technologies such as computers, cell phones and the internet, preventing studying from home [105].

Furthermore, since the impacts of the COVID-19 crises are more extreme to the most socioeconomically vulnerable populations, in addition to those located in regions with low or no access to basic needs, such as peripheral residents, indigenous people, women and children, the importance of intersectional thinking for gender equality is highlighted (SDG 5). There was already a large context of gender inequality and abuse before the pandemic crisis, but it is possible that lockdowns have worsened the problem [106].

At this point, it should be noted that the health labor market is also characterized by gender roles, with women representing around 70% of the health workforce [101, 106]. Nurses are more prone to exposure to the COVID-19 than doctors, since nurses and nursing technicians are in direct contact with secretions that spread of COVID-19 such as saliva, phlegm and feces.

Reference [107] remark that there is another vulnerable group that is even more affected by COVID-19: people with disabilities. This is because they have more intense care needs than others who fall ill with COVID-19, including longer hospital stays and more intensive nursing care. In addition, this group may need more sedation to deal with the hospital environment and, therefore, potentially have increased needs for intubation, which, for the most part, have not been met.

Regarding public security, some researchers like [104] point out that the moment marked by COVID-19 will cause surges in some types of criminal violence, while delaying others (for example, residential theft due to increased presence and domestic protection). In addition, the increase in domestic violence will ostensibly reflect the limits of social isolation, and that the justice apparatus will be challenged.

Regarding the prison population, Ref. [108] point out that these people are neglected in the COVID-19 scenario, even though it is known that overcrowding, poor hygiene and inadequate access to medical care make correctional facilities particularly vulnerable to the spread of the virus, many of the prevention strategies recommended by WHO are impossible to put in place. Infection control measures, such as social distancing, hand washing and lockdowns are limited in densely populated prisons, where most prisoners share cells and other community spaces. These places are also notoriously unhealthy, lack adequate ventilation or cleaning materials, and prisoners often have restricted access to soap and running water [109].

On the other hand, the pandemic presents an opportunity to accelerate the criminal justice reform that is already underway [110]. An urgent national response is needed, since prison staff and incarcerated populations are disproportionately infected by COVID-19 and because populations involved in justice face additional disparities that make them more vulnerable [104].

In the same sense, Ref. [82] argue that the 2030 Agenda presents the best possible approach to manage COVID-19 with the aim of ensuring that, now and in the future, human well-being is achieved and, at the same time, ecological and economic sustainability is preserved.

In addition, Ref. [13] point out that state and municipal governments in Brazil, within the scope of their constitutionally guaranteed competence in health matters, have addressed normative and administrative acts to restrict the movement of people, establish the mandatory isolation of individuals, and make determinations to perform diagnostic tests on individuals.

However, widespread discontent over current social arrangements—even among smaller population groups—can lead to a boycott and sabotage of implemented health measures and distrust of government officials. Therefore, it is crucial to adopt a broader understanding of pandemic preparedness as a public good and build social cohesion by meeting the demands of cooperative justice to encourage widespread cooperation and thus improve resilience to public health emergencies [111]. In this sense, Ref. [20] argues that due to the exponential multiplication of the virus, communities will remain vulnerable if they do not guarantee access to basic health and sanitation infrastructure for all.

Even before the pandemic, the world was far from meeting the SDGs [112], either because few efforts were being made to do so, or because in order to comply with the SDGs, the whole of society must be involved. That is, public policies are needed to connect areas such as environment, economy, politics, health, infrastructure, technology, gender equality, among others, so that the SDGs, in fact, are met [98].

In Brazil, associations such as the PAHO—the Pan American Health [113]—have supported the actions of the Ministry of Health of the Brazil in response to COVID-19 since January 2020, including the aforementioned emergency program that provided to vulnerable populations a monthly aid of R\$ 600 (six hundred reais). In March 2020, PAHO conducted training for public health specialists in Brazil in the use of Go.Data, a tool that seeks to facilitate the investigation of outbreaks and epidemics, such as the disease caused by the novel coronavirus. In addition, PAHO has purchased more than 10 million RT-PCR tests, which detect whether a person is infected with the novel coronavirus. Above all, PAHO has conducted a series of virtual seminars with specialists from different countries—including China, Spain, Italy and Japan—to support Brazil in the development of protocols, as well as to inform public health authorities.

Despite these efforts, the measures determined in Brazil and in the world were insufficient to curb the disease [109]. Furthermore, it is indisputable that COVID-19 and social isolation measures, by forcing families to stay at home to save lives and prevent the spread of the disease, exposed social inequalities and revealed not only the structural and historical inequality in the world, but the fragility of families with regard to access to current income and drinking water, which guarantees the consumption of essential goods [104, 114].

The future of post-pandemic humanity is still uncertain, either because there is no forecast of ending or decreasing the peak of the disease, or because there is no close precedent in recent history capable of drawing a parallel or even drawing similar ideas. This COVID-19 phenomenon is totally unique [98]. In the current context, falling income and rising unemployment, associated with insufficient emergency programs, will lead thousands of Brazilians to suffer from this crisis. Furthermore, the absence of more robust programs to protect families on the part of the Brazilian government and the flattening of the working class, associated with the increase in the extent of poverty, may imply an even longer way to achieve a more just society in post-pandemic times [109].

3.4   The Influence of Waste Generation by Covid-19 on Responsible Consumption and Production and on the Balance of the Environment

According to [17] COVID-19 was not a surprise, because new emerging infectious diseases were expected, mainly driven by the growth of human populations that increasingly disturb natural ecosystems. In addition, climate change is affecting factors such as the increased demand for animal protein, which is accompanied by viruses, bacteria and other pathogens that increase the likelihood of contracting zoonotic diseases, such as swine flu and avian flu.

Despite the great focus on health research, in the goal of mitigating the effects of the novel coronavirus, the impacts of this pandemic transcend the issues of body and mind [17, 115]. Through restrictions in Pandemic, such as social distancing, the behaviors of organizations, consumers, politicians and leaders in general have undergone major adaptations [4]. Commercial, manufacturing and mobility activities were limited [116], [117], many companies moved to the virtual work environment using the home office model, and unemployment rates increased. Such changes can contribute decisively to the construction of new social models, as well as to the organizations to manage risks and opportunities in this and in contexts of volatility and future uncertainties.

Some new practices and standards have promoted positive impacts, particularly on the environment [13]. Chinese researchers reaffirm the rapid decline in the levels of air pollution, accentuated in NO_2, given the lower recurrence of industrial production [118], in addition to the improvement in biodiversity and in tourist places. However, the challenging aspect of these changes lies in environmental sustainability practices. This new and complex moment has also significantly affected waste management [119].

The capitalist economic system causes society to present a posture of excessive, irresponsible and unsustainable consumption [4]. As a result, hospital and household waste ends up being disposed of in a way that is harmful to the environment. The volume of waste in the quarantine is exponentially increased, since people spend more time at home, and the face of an unexpected situation led to the intensification of the powers of consumption of energy, food and water [72], [117]. The Brazilian Association of Public Cleaning and Special Waste Companies (2021) reports that there was an increase of 15%–25% in the amount of residential waste, such as food, toilet paper, face masks, gloves, cleaning products and hand sanitizers [47, 117]. As for hospitals, the estimate is for a growth of 10–20 times in these materials.

Such products, such as facemasks and plastics, pose serious threats for the environment, given their lasting consequences for the planet—some of these products will last for about 450 years in nature. In Africa alone, in a study with a sample of 15 countries, 586,833,053 masks are discarded per day [120]. In Hong Kong there are reports that masks have accumulated on nature trails and beaches due to improper disposal in water courses directly affecting the biodiversity of the marine ecosystem [117].

A high volume of waste generation of hospital materials—such as gloves, aprons, masks and other protective clothing and equipment—contradicts the tendencies for environmental improvements, since the disposal of this waste is often being led to burning in the sky open air and incineration, a fact that can affect air quality and health implications due to exposure to toxins [121]. In addition, the COVID-19 pandemic generated unprecedented and worrying contamination in the oceans [14].

Several NGOs such as Oceans Asia; WWF and OpérationMerPropre have denounced the impact of this new type of waste on marine ecosystems. The destination of this waste to rivers and oceans is disruptive to marine life. Non-biodegradable plastics (polypropylene), present in masks for example, can be ingested by water animals and consequently cause acidification of the oceans.

In 2018, before the pandemic, the oceans and marine life had received about 13 million tonnes of plastic [122]. In the same period, Brazil was affected by the destruction of species and ecosystems to support human demand for animal protein, such as the deforestation of biologically rich Amazonian forests and the tropical savannah of the Cerrado in Brazil to graze cattle.

The explosive spread of the virus has not given countries enough time to adapt to this new situation. The lack of preparation of adequate protocols, the incorrect handling of increasing volumes of medical waste and deficiencies in the management of medical and domestic waste collection services can increase the medium and long-term levels of plastic pollution on beaches, coasts and rivers South America [82].

The new "normal" requires new theoretical approaches to action planning in situations of uncertainty, for the public, private and civil society [119]. Deficiencies in management systems must be addressed as a primary objective to achieve sustainable development and reduce the environmental impact caused by waste on the coasts and seas. In addition, it is extremely important to apply strategies and policies to achieve more sustainable and responsible coastal tourism, strengthening care and respect for the environment [82].

Although the focus is the environmental concern about the increase in the generation of waste, this indicator directly impacts the health of the population and indirectly the economy [82]. The greater number deaths has also meant less contributions to socioeconomic development. Waste management is also a priority for guaranteeing human well-being during the pandemic.

In the same vein, Ref. [121] highlight that given the high costs of sustainable solid waste management (SWM), governments are increasingly associating themselves with the private sector through public–private partnerships to find sustainable solutions, which can be an outlet for implementing joint post-pandemic recovery policies.

In addition, the socioeconomic crisis reshaped investment in energy and affected the sector significantly with disruptions due to mobility restrictions [103]. With energy, health, food, and economic insecurity, it remains a challenge for countries to intertwine in practice plans corresponding to the goals of the SDGs of an urgent nature.

As such, several countries have instituted policies to ensure sustainable waste management, protecting the safety of handlers of these materials as well as diminishing air, soil and water pollution. Although, COVID-19 has distracted governments and the public from many other environmental problems, those are still getting worse, especially the loss of biodiversity and the damaging ecological impacts of climate change. These interconnected problems inevitably intensified, unless a serious policy of care generates effective actions to solve them [17].

Some private organizations see this issue as an opportunity to generate income [103]. In Brazil, a startup that provided services for collecting residential and business organic leftovers for composting and organic fertilizers had a 25% increase in income.

The period marked by COVID-19 will serve as an example of a crisis that has raised unprecedented challenges for the loss and waste of food in the global food supply system, therefore testing our commitment to the principles adopted by Agenda 2030 by demonstrating that the food system that we built and as we know it today is flawed and vulnerable. Although the SDGs do not specifically state that there is a human right to food, SDG 2 envisions a better and more just world that is based on the sufficiency of the global food supply, environmental sustainability and food security for all [72].

SDG 12 addresses sustainable production and consumption, with a focus on global (12.1) and local actions, such as achieving efficient use of natural resources (12.2). In addition to that, the targets aim to reduce food waste (12.3), handling chemical waste responsibly (12.4), above all, managina solid waste (12.5) and reducing pollutant emissions (12.4). We need to rebalance the ecosystem, in all of its relations: with nature, in human relations, in the economy, in the productive processes and in their value chains (Institute of Applied Economic Research, 2021).

The main strategies in line with sustainable consumption in response to the pandemic are aimed at reducing consumption and strengthening and encouraging responsible production in the face of the prospect of a crisis in production and supply. On the other hand, less circulation of people can reduce superfluous consumption and disposal [18]. The goals and parameters of responsible consumption and production can guide the plans and protocols for protection and economic recovery from the crisis. One example is the concern of countries to maintain the provision of basic urban services—including collection and management of waste that is becoming a growing challenge for cities struggling with the consequences of COVID-19 [121].

Furthermore, the reduction of waste in the food logistics chains and the targeting of their surpluses, as well as in the health sector are also challenging [72]. Finally, incentives for recycling cooperatives are also necessary, as in Brazil, where they are responsible for a high rate of aluminum recycling in the country, and where it was verified that during the pandemic, recycling programs suffered falls [121].

The development of the COVID-19 pandemic is a consequence of urbanization and aggravated by some existing social and environmental problems: environmental degradation, flow of pathogens between humans and animals, inequalities and a lack of preventive social and public health measures [16]).

Although COVID-19 emerged from reservoirs of wild animals linked to environmental disturbances, transmission occurred through humans and facilitated by economic globalization [17]. While COVID-19 brought economic, environmental and social challenges, it also brought an opportunity to bring transformational changes to the structure and functioning of the global economy [100].

The essential links between human health, well-being, biodiversity and climate change can inspire a new generation of innovators to provide green solutions to enable humans to live in a healthy balance with nature, leading to a resilient future. The pandemic also shows itself as a unique opportunity in a generation to rethink how we grow our economies in a way that does not put the global environment at risk as we have done in recent decades [99]. The downsizing of the consumer economy and fundamental changes in global production networks and supply chains [115] will prompt a search for a new world economy with strategies to safeguard biodiversity and human well-being [4].

Finally, the systematization of the knowledge obtained in this study is represented in Table 1, establishing relationships between the SDGs and mitigating solutions for the impacts of this pandemic on sustainability.

4 Conclusions

The purpose of this chapter was to demonstrate how the impacts of the COVID-19 pandemic imply compliance with the SDGs. For this, we have used a literature review to support our arguments and categorize indicators of the relationships we aspire for. In this context, we identified the impacts of the novel coronavirus widespread in all spheres of sustainable development, with environmental, social, economic and governance aspects, as explained below.

Regarding sustainability in the governance sphere, COVID-19 showed the world that most countries were not prepared to deal with the virus's spread. Surprisingly, not even the most economically developed countries were able to implement efficient strategies to prevent contaminated people from entering and leaving the country, spread the virus, and collapse of health systems. In a globalized and highly connected world, institutions have failed to join forces to control COVID-19. At this point, the global partnership is essential to control the spread of the virus, research and technology for the development of vaccines, human trafficking.

While SDGs 16 and 17 are stimulating guides for promoting peace, justice, and cooperation between nations and institutions, we have identified critical social justice issues that need to be addressed to increase citizens' volunteerism and compliance with prevention and mitigation measures. In this context, the COVID-19 pandemic crisis may offer a rare and invaluable opportunity to rethink and redefine the economy towards a better and more sustainable future.

Humankind is facing an unprecedented crisis in terms of social sustainability that needs to cultivate global solidarity and concern for the well-being of all. Therefore, we highlight the importance of the social dimension of SDG2, SDG3, SDG4, SDG5 and SDG10, as they help to reshape our relationship with the natural environment. The integrative literature review conducted with a focus on COVID-19 and its

Table 1 SDGs, mitigating solutions for impacts on sustainability based on human actions

Impact	Negative implications	Mitigating solutions	Links with SDGs	Authors
Environmental	A—Increase in waste production; B—Increase in hospital waste; C—Reduction of recycling programs; D—Pollution of the seas; E—Air pollution F—Compromise of marine and terrestrial biodiversity;	1—Waste management; 2—Hygiene plan; 3—Responsible and conscious consumption; 4—Preservation of ecosystems based on the control of waste, mainly plastic; 5—Control of greenhouse gas emissions; 6—Awareness of water consumption	6 CLEAN WATER AND SANITATION 12 RESPONSIBLE CONSUMPTION AND PRODUCTION 13 CLIMATE ACTION 14 LIFE BELOW WATER 15 LIFE ON LAND	[1–4, 12–14, 45, 1, 17, 2, 23, 72, 3, 4, 47, 100, 115–119, 118]
Economic	A—Decrease in GDP; B—Unemployment C—Closing industries and businesses; D—Increase in the consumption of residential energy; E—Economic stress for formal and informal microentrepreneurs; F—Decrease in urban mobility; G—Economic decline; H—Increase in poverty, I—Authoritarianism, J—Increase in urban violence and crime K—Increase in social inequalities	1—Joint action between public and private; 2—Formulation of policies that best apply approaches combining economic recovery and sustainability; 3—Subsidies to low-income populations; 4—Aid and exemptions for small and medium-sized companies; 5—Investment in ICT for business adaptations; 6—Plans for displacement of people with intelligent mobility; 7—International cooperation between all international agents; 8—Better policies on labor rights; 9—Energy distribution	1 NO POVERTY 7 AFFORDABLE AND CLEAN ENERGY 8 DECENT WORK AND ECONOMIC GROWTH 9 INDUSTRY INNOVATION AND INFRASTRUCTURE 11 SUSTAINABLE CITIES AND COMMUNITIES	[116, 10, 70, 112, 95, 79, 99, 123, 19, 46, 84, 86, 94, 98, 97, 97–99, 83, 117, 20, 69]

(continued)

Table 1 (continued)

Impact	Negative implications	Mitigating solutions	Links with SDGs	Authors
Social	A—Physical health with mortality rate and lack of beds in hospitals; B—Mental health problems due to social isolation; C—Uneven impact on women, especially those who belong to blacks and indigenous populations, who are more impacted than men by less access to quality of life; D—Increase in domestic violence E—Increase in crime; F—Increase in social inequality and access to basic needs, such as housing, health systems, hygiene, basic sanitation and food; G—Difficulties in international trade cooperation, such as vaccinations; H—Decrease in access to education in underdeveloped countries due to the lack of access to the internet; I—Overcrowding in camps, settlements and shelters for groups more prone to crowding—homeless people and refugees; J—Overloading the public machine;	1—Investment in research and vaccination; 2—Emergency basic needs subsidy plans; 3—Actions aimed at gender equality; 4—Reduce inequality through public aid and policies that enable housing, education and health for all; 5—Strengthening democracies; 6—Strengthening international cooperation; 7—Education with remote education and subsidies for technological access for students and employees; like chips, internet and videos on free platforms 8—Welcoming actions for homeless people and refugees; 9—Control the traffic of people between countries and states during and after the pandemic; 10—Inclusion of family planning and reproductive health services in the countries; essential services package; 11—Creation of a global fund with the objective of strengthening health systems, so that they are able to face new global pandemics, at the same time that such fund would contribute to the achievement of the SDG3	2 ZERO HUNGER 3 GOOD HEALTH AND WELL-BEING 4 QUALITY EDUCATION 5 GENDER EQUALITY 10 REDUCED INEQUALITIES 16 PEACE, JUSTICE AND STRONG INSTITUTIONS 17 PARTNERSHIPS FOR THE GOALS	[56, 27, 7–9, 65, 66, 17–21, 19, 33, 46, 61, 111, 97, 87, 32, 28, 55–57, 60–66, 57, 3, 62, 4, 60, 101, 40, 101–104, 106–112, 96]; Sarricolea (2020), Lopez (2021), and Siddiqui et al. (2020)

impacts on the social, economic and political inequalities of the world's populations demonstrated that the central hypothesis accepted by the theory is that epidemic events have the effect of reducing inequalities in the face of mortality of the poorest and the consequent revaluation of work due to the scarcity of labor.

It is undeniable that individuals and groups previously harmed by the social injustice created by the uneven distribution of power, money, and resources were even more affected by the pandemic. In this respect, we highlight women—whether due to contamination or due to the increase in domestic violence and crime rates, less access to energy, water, and food; children, faced with reduced access to education in underdeveloped countries due to the lack of internet access, to refugees, due to overcrowding in camps, settlements and shelters for groups more prone to crowding; the elderly, behold, they are more likely to get the disease and; especially the poorest, behold, simple measures to prevent the spread of new coronavirus, such as frequent hand washing and social distance, are not available to millions of people worldwide, as well as because disadvantaged groups are most directly affected and disproportionate due to existing poor health.

Specifically, this research's findings demonstrate that, at least within the limitations of the defined search scope, despite the reasonable number of articles found, no studies with a focus or significant discussion on targets 3.5, 3.6, 3.7, and 3.9 and were filtered out. If it is not possible to affirm with this research alone that these SDG 3 goals have not yet been an object of significant study by the academy, it is possible to indicate that studies focused on COVID-19 impact on these specific goals will need further research.

In addition, in the case of SDG 3, in the integrative review of the literature and in the analysis of the content of the articles relating the COVID-19 outbreak to the SDG 3, the findings reported that most articles maintain that the impacts of the COVID-19 pandemic go well beyond those issues regarding health. Finally, as a suggestion for future studies, we recommend conducting research aimed at understanding the impacts of COVID-19 on each of the SDG3 targets.

On the other hand, the search for articles in the gray literature and in the databases showed as a result that the central hypothesis discussed by various authors is that the COVID-19 pandemic tends to deepen socioeconomic inequality throughout the world due mainly to effects of restrictions on economic activity, a theory that is corroborated by [124] and the other authors of the commission formed by The Lancet, perhaps the journal with the most significant impact on health issues, for COVID-19.

In this context, all the scientific articles found postulate the same hypothesis of increasing inequalities and adopting, without exception, a transversal time frame, whereby the long-term effects of the pandemic are yet to be observed. Thus, even with this amount of evidence in the literature supporting the hypothesis that the COVID-19 pandemic should widen economic inequalities, as this is still an ongoing phenomenon, it is not possible to definitively rule out the opposite hypothesis.

In 2020, the pandemic had an overwhelming influence on the decline in GDP, especially in underdeveloped countries. This is because many industries, businesses, and new investments in non-essential sectors have been forced to cease or shrink their production. This reality was reflected in the increase in unemployment rates, falls in the stock exchanges, not to mention the economic stress suffered by formal and informal micro-entrepreneurs. On the other hand, other ventures such as supermarkets and technology companies have managed to survive. In this context, the mitigating solutions to such issues must be based on the integration of government incentive actions for poverty control and contemplating SDG1. In addition, it is necessary to remodel companies, universities, and institutions in general, focusing on innovations and in SDG 7, SDG8, SDG 9, SDG 11.

Finally, regarding environmental sustainability, the works used as references demonstrated an unprecedented impact on the environment with the increase of waste, hospital waste, reduction of recycling programs, increased pollution of the seas and air, and compromise of marine biodiversity terrestrial. These environmental problems are directly reflected in SDG 6, SDG 12, SDG 13, SDG 14 and SDG 15.

In order to minimize the impact on the environment, there was a need to rebalance the ecosystem in all its relations: with nature, in human relations, in the economy, in production processes and their value chains. In environmental terms, above all, responsible consumption and production (SDG 12) are understood to be the key to economic and environmental recovery from the crisis. Also, actions such as preserving ecosystems, controlling waste, hygiene plans, and waste management were considered mitigating solutions that can mitigate solutions for the impacts of COVID-19 on the environment.

The COVID-19 pandemic reinforces the need for social inclusion measures and guarantees the dignity of the human person. This critical moment's impacts are difficult to estimate in the medium and long-term, giving space to speculation and uncertainty. Although the origins of COVID-19 have also been discussed, the spread of this virus has highlighted the health risks that certain types of food products of animal origin present. In this sense, the coronavirus outbreak served to demonstrate, once again, the links between habitat loss and biodiversity and the increasing likelihood that infections will spread from wildlife to humans as zoonotic diseases.

The COVID-19 pandemic puts pressure on the SDGs' actions and introduces immediate needs that cannot wait until 2030 to be addressed. Therefore, the global objectives can be understood as a tool to promote actors' mobilization in societies—including government, corporate and civil society agents—to promote all its dimensions. Finally, international cooperation plays a fundamental role in development in the context of crisis, since the pandemic has shown that one can no longer think of isolated solutions, but of solutions that, respecting each country's particularity, can be sustainable. Sustainable as an interdependent goal.

In short, the chapter demonstrates that the SDGs, while suffering a severe setback with the pandemic, is a way of fighting the pandemic and, equally, of guiding the government, corporations and civil society towards the reconstruction of a genuine new normal committed to the constitutional principle of the dignity of the human person and the protection of the environment as a fundamental right of emerging generations.

Acknowledgements This chapter was conducted by the Centre for Sustainable Development (Greens), from the University of Southern Santa Catarina (Unisul) and Ânima Institute - AI, in the context of the project BRIDGE—Building Resilience in a Dynamic Global Economy: Complexity across scales in the Brazilian Food-Water-Energy Nexus; funded by the Newton Fund, Fundação de Amparo à Pesquisa e Inovação do Estado de Santa Catarina (FAPESC), Coordenação de Aperfeiçoamento de Pessoal de Nível superior (CAPES), National Council for Scientific and Technological Development (CNPq) and the Research Councils United Kingdom (RCUK).

References

1. López-Feldman A, Chávez C, Vélez MA, Bejarano H, Chimeli AB, Féres J, Robalino J, Salcedo R, Viteri C (2020) COVID-19: impacts on the environment and the achievement of the SDGS in Latin America. Desarrollo y Sociedad 2020(86). https://doi.org/10.13043/DYS.86.4
2. Muhammad S, Long X, Salman M (2020) COVID-19 pandemic and environmental pollution: A blessing in disguise? Sci Total Environ 728:138820
3. Ozili PK, Arun T (2020) Spillover of COVID-19: impact on the global economy. https://doi.org/10.2139/ssrn.3562570
4. Reicher S, Stott C (2020) On order and disorder during the COVID-19 pandemic. Br J Social Psychol 59(3):694–702. https://doi.org/10.1111/bjso.12398
5. Alfani G (2020) Epidemics, inequality and poverty in preindustrial and early industrial times. J Econ Lit (forthcoming)
6. Patel MS, Phillips CB (2021) COVID-19 and the moral imagination. The Lancet 397(10275):648–650. https://doi.org/10.1016/S0140-6736(21)00151-3
7. Watson MF, Bacigalupe G, Daneshpour M, Han WJ, Parra-Cardona R (2020) COVID-19 interconnectedness: health inequity, the climate crisis, and collective trauma. Fam Process 59(3):832–846. https://doi.org/10.1111/famp.12572
8. Santos BS (2020) A Cruel Pedagogia do Vírus. Almedina, Lisboa
9. Dimaggio AR (2020) Unequal America: class conflict, the news media, and ideology in an era of record inequality. Routledge, London
10. Berchin II, de Andrade JBSO (2020) GAIA 3.0: effects of the coronavirus disease 2019 (COVID-19) outbreak on sustainable development and future perspectives. Res Glob 2:100014.
11. United Nations (2020a) Shared Responsibility, Global Solidarity: Responding to the socio-economic impacts of COVID-19. https://unsdg.un.org/sites/default/files/2020-03/SG-Report-Socio-Economic-Impact-of-Covid19.pdf.
12. Chan AHY, Rutter V, Ashiru-Oredope D, Tuck C, Babar ZUD (2020) Together we unite: the role of the Commonwealth in achieving universal health coverage through pharmaceutical care amidst the COVID-19 pandemic. J Pharm Policy Pract 13:1–7. https://doi.org/10.1186/s40545-020-00214-6
13. Severo EA, De Guimarães JCF, Dellarmelin ML (2021) Impact of the COVID-19 pandemic on environmental awareness, sustainable consumption and social responsibility: Evidence from generations in Brazil and Portugal. J Clean Prod 286:124947. https://doi.org/10.1016/j.jclepro.2020.124947
14. Ardusso M, Forero-López AD, Buzzi NS, Spetter CV, Fernández-Severini MD (2021) COVID-19 pandemic repercussions on plastic and antiviral polymeric textile causing pollution on beaches and coasts of South America. Sci Total Environ 763:144365
15. World Health Organization (2021) Weekly operational update on COVID-19.1 March 2021. https://www.who.int/publications/m/item/weekly-operational-update-on-covid-19---1-march-2021
16. Rupani PF, Nilashi M, Abumalloh RA, Asadi S, Samad S, Wang S (2020) Coronavirus pandemic (COVID-19) and its natural environmental impacts. Int J Environ Sci Technol 17:1–12. https://doi.org/10.1007/s13762-020-02910-x

17. McNeely JA (2021) Nature and COVID-19: the pandemic, the environment, and the way ahead. Ambio 50(4):767–781. https://doi.org/10.1007/s13280-020-01447-0
18. Bebbington J, Unerman J (2020) Advancing research into accounting and the UN sustainable development goals. Acc Auditing Account J 33(7):1657–1670. https://doi.org/10.1108/AAAJ-05-2020-4556
19. Elavarasan RM, Pugazhendhi R (2020) Restructured society and environment: A review on potential technological strategies to control the COVID-19 pandemic. Sci Total Environ 725:138858
20. Timmermann C (2020) Pandemic preparedness and cooperative justice. Dev World Bioethics. https://doi.org/10.1111/dewb.12289
21. Wilson K (2020) The COVID-19 pandemic and the human rights of persons with mental and cognitive impairments subject to coercive powers in Australia. Int J Law Psychiatry 73:101605. https://doi.org/10.1016/j.ijlp.2020.101605
22. Beigi S (2020) How do the COVID-19 prevention measures interact with sustainable development goals? Preprint at https://doi.org/10.20944/preprints202010.0279.v1.
23. Munasinghe M (2020) COVID-19 and sustainable development. Int J Sustain Dev 23(1/2):1–24. https://doi.org/10.1504/IJSD.2020.112182
24. Roberton T, Carter ED, Chou VB, Stegmuller AR, Jackson BD, Tam Y, Sawadogo-Lewis T, Walker N (2020) Early estimates of the indirect effects of the COVID-19 pandemic on maternal and child mortality in low-income and middle-income countries: a modelling study. Lancet Glob Health 8(7):e901–e908
25. Mendenhall E (2020) The COVID-19 syndemicis not global: context matters. The Lancet 396(10264):1731. https://doi.org/10.1016/S0140-6736(20)32218-2
26. Horton R (2020) Offline: COVID-19 is not a pandemic. The Lancet 396(10255):874. https://doi.org/10.1016/S0140-6736(20)32000-6
27. Adebisi YA, Alaran AJ, Akinokun RT, Micheal AI, Ilesanmi EB, Lucero-Prisno DE III (2020) Sex workers should not be forgotten in Africa's COVID-19 response. Am J Trop Med Hyg 103(5):1780–1782
28. Leal Filho W, Brandli LL, Lange Salvia A, Rayman-Bacchus L, Platje J (2020) COVID-19 and the UN sustainable development goals: threat to solidarity or an opportunity? Sustainability 12(13):5343. https://doi.org/10.3390/su12135343
29. Clay JM, Parker MO (2020) Alcohol use and misuse during the COVID-19 pandemic: a potential public health crisis? Lancet Public Health 5(5):e259. https://doi.org/10.1016/S2468-2667(20)30088-8
30. Capasso A, Jones AM, Ali SH, Foreman J, Tozan Y, Di Clemente RJ (2021) Increased alcohol use during the COVID-19 pandemic: The effect of mental health and age in a cross-sectional sample of social media users in the US. Prev Med 145:106422. https://doi.org/10.1016/j.ypmed.2021.106422
31. Campbell E, Zahoor U, Payne A, Popova D, Welman T, Pahal GS, Sadigh P (2020) The COVID-19 pandemic: the effect on open lower limb fractures in a London major trauma centre-a plastic surgery perspective. Injury 52(3):402–406. https://doi.org/10.1016/j.injury.2020.11.047
32. Kumar N (2020) COVID 19 era: a beginning of upsurge in unwanted pregnancies, unmet need for contraception and other women related issues. Eur J Contracept Reproductive Health Care 25(4):323–325. https://doi.org/10.1080/13625187.2020.1777398
33. Ferreira-Filho ES, de Melo NR, Sorpreso ICE, Bahamondes L, Simões RDS, Soares-Júnior JM, Baracat EC (2020) Contraception and reproductive planning during the COVID-19 pandemic. Expert Rev Clin Pharmacol 13(6):615–622. https://doi.org/10.1080/17512433.2020.1782738
34. Sharma V, De Beni D, Sachs Robertson A, Maurizio F (2020) Why the promotion of family planning makes more sense now than ever before?. J Health Manag 22(2):206–214. https://doi.org/10.1177/0972063420935545
35. Organisation for Economic Co-operation and Development (2020) COVID-19 crisis in the MENA region: impact on gender equality and policy responses. Organisation for Economic Co-Operation and Development, Paris, pp 1–38.

36. Tang K, Gaoshan J, Ahonsi B, Ali M, Bonet M, Broutet N, Kara E, Kim C, Thorson A, Thwin SS (2020) Sexual and reproductive health (SRH): a key issue in the emergency response to the coronavirus disease (COVID- 19) outbreak. Reprod Health 17(1):17–19. https://doi.org/10.1186/s12978-020-00926-7

37. Otu A, Danhoundo G, Yaya S (2021) Prioritizing sexual and reproductive health in the face of competing health needs: where are we going? Reprod Health 18(1):2021–2024. https://doi.org/10.1186/s12978-021-01068-0

38. Dasgupta A, Kantorová V, Ueffing P (2020) The impact of the COVID-19 crisis on meeting needs for family planning: a global scenario by contraceptive methods used. Gates Open Res 4:102. https://doi.org/10.12688/gatesopenres.13148.2

39. Patel DP, Punjani N, Guo J, Alukal JP, Li PS, Hotaling JM (2021) The impact of SARS-CoV-2 and COVID-19 on male reproduction and men's health. Fertil Steril 115(4):813–823. https://doi.org/10.1016/j.fertnstert.2020.12.033

40. Rivillas-García JC, Cifuentes-Avellaneda Á, Ariza-Abril JS, Sánchez-Molano M, Rivera-Montero D (2021) Venezuelan migrants and access to contraception in Colombia: a mixed research approach towards understanding patterns of inequality. J Migr Health 3:100027. https://doi.org/10.1016/j.jmh.2020.100027

41. Lin TK, Law R, Beaman J, Foster DG (2021) The impact of the COVID-19 pandemic on economic security and pregnancy intentions among people at risk of pregnancy. Contraception 103(6):380–385. https://doi.org/10.1016/j.contraception.2021.02.001

42. Blanchet K, Alwan A, Antoine C, Cros MJ, Feroz F, Guracha TA, Haaland O, Hailu A, Hangoma P, Jamison D, Memirie ST, Miljeteig I, Naeem AJ, Nam SL, Norheim OF, Verguet S, Watkins D, Johansson KA (2020) Protecting essential health services in low-income and middle-income countries and humanitarian settings while responding to the COVID-19 pandemic. BMJ Global Health 5(10):e003675

43. Eissa N (2020) Pandemic preparedness and public health expenditure. Economies 8(3):60. https://doi.org/10.3390/economies8030060

44. Tediosi F, Lönnroth K, Pablos-Méndez A, Raviglione M (2020) Build back stronger universal health coverage systems after the COVID-19 pandemic: the need for better governance and linkage with universal social protection. BMJ Global Health 5(10):e004020

45. Lal A, Erondu NA, Heymann DL, Gitahi G, Yates R (2020) Fragmented health systems in COVID-19: rectifying the misalignment between global health security and universal health coverage. The Lancet 397(10268):61–67. https://doi.org/10.1016/S0140-6736(20)32228-5

46. Friedman EA, Gostin LO, Maleche A, Nilo A, Foguito F, Rugege U, Stevenson S, Gitahi G, Ruano AL, Barry M, Hossain S, Lucien F, Rusike I, Hevia M, Alwan A, Cameron E, Farmer P, Flores W, Hassim A, Mburu R, Mukherjee J, Mulumba M, Puras D, Periago R (2020) Global Health in the age of COVID-19: responsive health systems through a right to health fund. Health Hum Rights 22(1):199. https://doi.org/10.1186/s12936-020-03488-y

47. Rume T, Islam SDU (2020) Environmental effects of COVID-19 pandemic and potential strategies of sustainability. Heliyon 6(9):e04965

48. Dharmaraj S, Ashokkumar V, Hariharan S, Manibharathi A, Show PL, Chong CT, Ngam-charussrivichai C (2021) The COVID-19 pandemic face mask waste: a blooming threat to the marine environment. Chemosphere 272:129601

49. Garg S, Deshmukh C (2020) Tobacco: an invisible and immediate threat for Covid 19. Indian J Community Health 32(2):248–250. https://doi.org/10.47203/ijch.2020.v32i02supp.014

50. González-Marrón A, Martínez-Sánchez JM (2020) Correlation between prevalence of tobacco smoking and risk and severity of COVID-19 at the national level in the European Union: An ecological study. MedRxiv 2020:1–8. https://doi.org/10.1101/2020.04.28.20083352

51. Gupta AK, Nethan ST, Mehrotra R (2021) Tobacco use as a well-recognized cause of severe COVID-19 manifestations. Res Med 176:106233. https://doi.org/10.1016/j.rmed.2020.106233

52. Magfira N, Helda H (2020) Correlation between adult tobacco smoking prevalence and mortality of Coronavirus Disease-19 across the world. MedRxiv 1–14. https://doi.org/10.1101/2020.12.01.20241596

53. Clancy L, Gallus S, Leung J, Egbe CO (2021) Tobacco and COVID-19: understanding the science and policy implications. Tobacco Induced Dis, 18:1–4. https://doi.org/10.18332/TID/131035

54. Alla F, Berlin I, Nguyen-Thanh V, Guignard R, Pasquereau A, Quelet S, Schwarzinger M, Arwidson P (2020) Tobacco and COVID-19: a crisis within a crisis? Can J Pub Health 111(6):995–999. https://doi.org/10.17269/s41997-020-00427-x

55. Djalante R, Shaw R, De Wit A (2020) Building resilience against biological hazards and pandemics: COVID-19 and its implications for the Sendai Framework. Prog Disaster Sci 6:100080

56. Addo-Atuah J, Senhaji-Tomza B, Ray D, Basu P, Loh FHE, Owusu-Daaku F (2020) Global health research partnerships in the context of the Sustainable Development Goals (SDGs). Res Social Adm Pharm 16(11):1614–1618

57. Osingada CP, Porta CM (2020) Nursing and Sustainable Development Goals (SDGs) in a COVID-19 world: the state of the science and a call for nursing to lead. Public Health Nursing 37(5):799–805. https://doi.org/10.1111/phn.12776

58. Bukhari N, Manzoor M, Rasheed H, Nayyer B, Malik M, Babar ZUD (2020) A step towards gender equity to strengthen the pharmaceutical workforce during COVID-19. J Pharm Policy Pract 13:1–5. https://doi.org/10.1186/s40545-020-00215-5

59. Rosa WE, Gray TF, Chow K, Davidson PM, Dionne-Odom JN, Karanja V, Khanyola J, Kpoeh JDN, Lusaka J, Matula ST, Mazanec P, Moreland PJ, Pandey S, de Campos AP, Meghani SH (2020) Recommendations to leverage the palliative nursing role during COVID-19 and future public health crises. J Hosp Palliat Nurs. 22(4):260–269. https://doi.org/10.1097/NJH.0000000000000665

60. Reis RF, de Melo Quintela B, de Oliveira Campos J, Gomes JM, Rocha BM, Lobosco M, Dos Santos RW (2020) Characterization of the COVID-19 pandemic and the impact of uncertainties, mitigation strategies, and underreporting of cases in South Korea, Italy, and Brazil. Chaos, Solitons Fractals 136:109888

61. Golar G, Malik A, Muis H, Herman A, Nurudin N, Lukman L (2020) The social-economic impact of COVID-19 pandemic: implications for potential forest degradation. Heliyon 6(10):e05354

62. Pereira M, Oliveira AM (2020) Poverty and food insecurity may increase as the threat of COVID-19 spreads. Public Health Nutr 23(17):3236–3240. https://doi.org/10.1017/S136898002000349 3

63. Waller R, Hodge S, Holford J, Milana M, Webb S (2020) Lifelong education, social inequality and the COVID-19 health pandemic. Int J Lifelong Educ 39(3):243–246. https://doi.org/10.1080/02601370.2020.1790267

64. Ashford NA, Hall RP, Arango-Quiroga J, Metaxas KA, Showalter AL (2020) Addressing inequality: the first step beyond COVID-19 and towards sustainability. Sustainability 12(13):5404

65. Blundell R, Costa Dias M, Joyce R, Xu X (2020) COVID-19 and Inequalities. Fisc Stud 41(2):291–319. https://doi.org/10.1111/1475-5890.12232

66. Brennen JS, Simon F, Howard PN, Nielsen RK (2020) Types, sources, and claims of COVID-19 misinformation. Reuters Institute 7:1–3

67. Child J (2020) Organizational participation in post-covid society–its contributions and enabling conditions. Int Rev Appl Econ 35(2):1–30.https://doi.org/10.1080/02692171.2020.1774976

68. Makhoul J, Kabakian-Khasholian T, Chaiban L (2021) Analyzing the social context of health information and misinformation during the COVID-19 pandemic: a case of emerging inequities in Lebanon. Glob Health Promot 28(1):33–41

69. Bavel JV, Baicker K, Boggio PS, Capraro V, Cichocka A, Cikara M, Crockett MJ, Crum AJ, Douglas KM, Druckman JN, Drury JN (2020) Using social and behavioural science to support COVID-19 pandemic response. Nat Hum Behav 4(5):460–471. https://doi.org/10.1038/s41562-020-0884-z

70. Bruin WB, Saw HW, Goldman DP (2020) Political polarization in US residents' COVID-19 risk perceptions, policy preferences, and protective behaviors. J Risk Uncertain 61(2):177–194. https://doi.org/10.1007/s11166-020-09336-3
71. Lopez PJ, Neely AH (2021) Fundamentally uncaring: the differential multi-scalar impacts of COVID-19 in the US. Soc Sci Med 272:113707. https://doi.org/10.1016/j.socscimed.2021.113707
72. O'Hara S, Toussaint EC (2021) Food access in crisis: food security and COVID-19. Ecol Econ 180:106859
73. Hasen RL (2020) Three pathologies of American voting rights illuminated by the COVID-19 pandemic, and how to treat and cure them. Election Law J Rules Politics Policy 19(3):263–288. https://doi.org/10.1089/elj.2020.0646
74. de Carvalho CA, Viola PCDAF, Sperandio N (2021) How is Brazil facing the crisis of Food and Nutrition Security during the COVID-19 pandemic? Public Health Nutr 24(3):561–564. https://doi.org/10.1017/S1368980020003973
75. UNICEF (2020) Impactos Primários e Secundários da COVID-19 em Crianças e Adolescentes Relatório de análise 1ª Onda. https://www.unicef.org/brazil/media/11331/file/relatorio-analise-impactos-primarios-e-secundarios-da-covid-19-em-criancas-e-adolescentes.pdf
76. Kousi T, Mitsi LC, Simos J (2021) The early stage of COVID-19 outbreak in Greece: a review of the national response and the socioeconomic impact. Int J Environ Res Public Health 18(1):322
77. Baca G (2020) Eastern surveillance, Western malaise, and South Korea's COVID-19 response: oligarchic power in Hell Joseon. Dialect Anthropol 44(3):301–307
78. Ali TO, Hassan M, Hossain N (2021) The moral and political economy of the pandemic in Bangladesh: Weak states and strong societies during Covid-19. World Dev 137:105216
79. Colombo E (2020) Human Rights-inspired Governmentality: COVID-19 through a Human Dignity Perspective. Crit Sociol 47(4–5):571–581
80. Cimini F, Julião N, de Souza A, Cabral N (2020) Covid-19 pandemic, social mitigation and taxation: the open veins of inequality in Latin America. Bull Lat Am Res 39:56–61. https://doi.org/10.1111/blar.13213
81. Farris M, Sarricolea Espinoza P (2020) Le geografie della pandemia nel Cile della rivolta sociale: tra centralismo politico e vulnerabilità delle regioni rurali. Documenti Geografici 1:683–697.https://doi.org/10.19246/docugeo2281-7549/202001_43
82. Siddique A, Shahzad A, Lawler J, Mahmoud KA, Lee DS, Ali N, Bilal M, Rasool K (2021) Unprecedented environmental and energy impacts and challenges of COVID-19 pandemic. Environmental Research 193:110443
83. Salvati E (2021) Crisis and intergovernmental retrenchment in the European Union? Framing the EU's answer to the COVID-19 pandemic. Chin Polit Sci Rev 6(1):1–19. https://doi.org/10.1007/s41111-020-00171-0
84. Goniewicz K, Khorram-Manesh A, Hertelendy AJ, Goniewicz M, Naylor K, Burkle FM (2020) Current response and management decisions of the European Union to the COVID-19 outbreak: a review. Sustainability 12(9):3838. https://doi.org/10.3390/su12093838
85. Hamamra B, Alawi N, Daragmeh AK (2020) Covid-19 and the decolonisation of education in Palestinian universities. Educ Philos Theory 2020:1–15. https://doi.org/10.1080/00131857.2020.1865921
86. Haer R, Demarest L (2020) COVID-19 in Africa: Turning a Health Crisis into a Human Security Threat? Peace Econ Peace Sci Public Policy 26(3):1. https://doi.org/10.1515/peps-2020-0032
87. Kassa MD, Grace JM (2020) Race against death or starvation? COVID-19 and its impact on African populations. Public Health Rev 41(1):1–17. https://doi.org/10.1186/s40985-020-00139-0
88. Andrias K, Sachs BI (2021) Constructing countervailing power: law and organizing in an era of political inequality. Yale Law J 130(3):546–635
89. Méndez M, Flores-Haro G, Zucker L (2020) The (in) visible victims of disaster: Understanding the vulnerability of undocumented Latino/a and indigenous immigrants. Geoforum 116:50–62. https://doi.org/10.1016/j.geoforum.2020.07.007

90. Wannamakok W, Sissokho O, Gates TG (2020) Human rights and education for Gambian young women during COVID-19: recommendations for social policy and practice. Int Social Work 63(6):825–829. https://doi.org/10.1177/0020872820952860
91. Jaiswal J, LoSchiavo C, Perlman DC (2020) Disinformation, misinformation, and inequality-driven mistrust in the time of COVID-19: lessons unlearned from AIDS denialism. AIDS Behav 24:2776–2780. https://doi.org/10.1007/s10461-020-02925-y
92. Miconi D, Li ZY, Frounfelker RL, Venkatesh V, Rousseau C (2021) Socio-cultural correlates of self-reported experiences of discrimination related to COVID-19 in a culturally diverse sample of Canadian adults. Int J Intercult Relat 81:176–192. https://doi.org/10.1016/j.ijintrel.2021.01.013
93. Ralls D (2020) Beyond the rainbows: the missing voices of children and young people in this pandemic. *LSE Covid 19 Blog.*
94. Igwe PA, Ochinanwata C, Ochinanwata N, Adeyeye JO, Ikpor IM, Nwakpu SE, Egbo OP, Onyishi IE, Vincent O, Nwekpa KC Nwakpu KO (2020) Solidarity and social behaviour: how did this help communities to manage COVID-19 pandemic? Int J Soc Social Policy 40(9/10):1183–1200. https://doi.org/10.1108/IJSSP-07-2020-0276
95. Chapman CM, Miller DS (2020) From metaphor to militarized response: the social implications of "we are at war with COVID-19"–crisis, disasters, and pandemics yet to come. Int J Sociol Social Policy 40(9/10):1107–1124. https://doi.org/10.1108/IJSSP-05-2020-0163
96. Xafis V (2020) What's inconvenient for you, is life-saving for me": How health inequities are playing out during the COVID-19 pandemic. Asian Bioethics Rev 12(2):1–12. https://doi.org/10.1007/s41649-020-00119-1
97. Karaye IM, Horney JA (2020) The impact of social vulnerability on COVID-19 in the US: an analysis of spatially varying relationships. Am J Preventive Med 59(3):317–325. https://doi.org/10.1016/j.amepre.2020.06.006
98. Iwuoha JC, Jude-Iwuoha AU (2020) COVID-19: challenge to SDG and globalization. Electron Res J Social Sci Humanities 2(3):1–5
99. Dos Santos JLG, Stein Messetti PA, Adami F, Bezerra IMP, Maia PCG, Tristan-Cheever E, Abreu LCD (2021) Collision of fundamental human rights and the right to health access during the novel coronavirus pandemic. Front Public Health 8:553
100. Sarkar P, Debnath N, Reang D (2021) Coupled human-environment system amid COVID-19 crisis: A conceptual model to understand the nexus. Sci Total Environ 753:141757
101. Rhodes R (2020) Justice and guidance for the COVID-19 pandemic. Am J Bioethics 20(7):163–166. https://doi.org/10.1080/15265161.2020.1777354
102. van Zanten JA, van Tulder R (2020) Beyond COVID-19: Applying "SDG logics" for resilient transformations. J Int Bus Policy 3(4):451–464. https://doi.org/10.1057/s42214-020-00076-4
103. Mofijur M, Fattah IR, Alam MA, Islam AS, Ong HC, Rahman SA, Najafi G, Ahmed SF, Uddin MA, Mahlia TMI (2020) Impact of COVID-19 on the social, economic, environmental and energy domains: Lessons learnt from a global pandemic. Sustain Prod Consumption 26:343–359. https://doi.org/10.1016/j.spc.2020.10.016
104. Miller JM, Blumstein A (2020) Crime, justice & the CoViD-19 pandemic: toward a national research agenda. Am J Criminal Justice 45(4):515–524. https://doi.org/10.1007/s12103-020-09555-z
105. United Nations (2020b) Policy Brief: Education during COVID-19 and beyond. https://www.un.org/development/desa/dspd/wp-ontent/uploads/sites/22/2020/08/sg_policy_brief_covid-19_and_education_august_2020.pdf
106. Swan E (2020) COVID-19 foodwork, race, gender, class and food justice: an intersectional feminist analysis. Gender Manage Int J 35(7/8):693–703
107. Nelson RH, Francis LP (2020) Justice and intellectual disability in a pandemic. Kennedy Inst Ethics J 30(3):319–338
108. Mukherjee TI, El-Bassel N (2020) The perfect storm: COVID-19, mass incarceration and the opioid epidemic. Int J Drug Policy 83:102819. https://doi.org/10.1016/j.drugpo.2020.102819
109. Costa JSD, Silva JCFD, Brandão ESC, Bicalho PPG (2020) COVID-19 no sistema prisional brasileiro: da indiferença como política à política de morte. Psicologia Sociedade, 32.https://doi.org/10.1590/1807-0310/2020v32240218

110. Bester JC (2020) Justice, well-being, and civic duty in the age of a pandemic: why we all need to do our bit. J Bioethical Inquiry 17(4):737–742
111. Higgins-Desbiolles F (2020) Socialising tourism for social and ecological justice after COVID-19. Tourism Geogr 22(3):610–623.https://doi.org/10.1080/14616688.2020.1757748
112. Burke S (2020) Stronger together? Intergenerational connection and Covid-19. Qual Ageing Older Adults 21(4):253–259. https://doi.org/10.1108/QAOA-07-2020-0033
113. Pan American Health Organization (2021) Folha informativa COVID-19: Escritório da OPAS e da OMS no Brasil. https://www.paho.org/pt/covid19
114. Martins LD, da Silva I, Batista WV, de Fátima Andrade M, de Freitas ED, Martins JA (2020) How socio-economic and atmospheric variables impact COVID-19 and influenza outbreaks in tropical and subtropical regions of Brazil. Environ Res 191:110184
115. Casado-Aranda LA, Sánchez-Fernández J, Viedma-del-Jesús MI (2021) Analysis of the scientific production of the effect of COVID-19 on the environment: A bibliometric study. Environmental Research 193:110416
116. Anholon R, Rampasso IS, Silva DA, Leal Filho W, Quelhas OLG (2020) The COVID-19 pandemic and the growing need to train engineers aligned to the sustainable development goals. Int J Sustain High Educ 21(6):1269–1275. https://doi.org/10.1108/IJSHE-06-2020-0217
117. Sarkodie SA, Owusu PA (2020) Impact of COVID-19 pandemic on waste management. Environ Dev Sustain 23:7951–7960. https://doi.org/10.1007/s10668-020-00956-y
118. Wang Q, Su M (2020) A preliminary assessment of the impact of COVID-19 on environment: a case study of China. Sci Total Environ 728:138915
119. Sarkis J (2020) Supply chain sustainability: learning from the COVID-19 pandemic. Int J Oper Prod Manag 41(1):63–73. https://doi.org/10.1108/IJOPM-08-2020-0568
120. Nzediegwu C, Chang SX (2020) Improper solid waste management increases potential for COVID-19 spread in developing countries. Resour Conserv Recycl 161:104947. https://doi.org/10.1016/j.resconrec.2020.104947
121. Urban RC, Nakada LYK (2021) COVID-19 pandemic: Solid waste and environmental impacts in Brazil. Sci Total Environ 755:142471
122. United Nations (2019) Clima e meioambiente. https://news.un.org/pt/story/2019/06/1675231
123. Dutta A, Fischer HW (2021) The local governance of COVID-19: Disease prevention and social security in rural India. World Dev 138:105234
124. Sachs JD, Karim SA, Aknin L, Allen J, Brosbøl K, Barron GC, Daszak P, Espinosa MF, Gaspar V, Gaviria A, Bartels JG (2020) Lancet COVID-19 Commission Statement on the occasion of the 75th session of the UN General Assembly. The Lancet 396(10257):1102–1124. https://doi.org/10.1016/S0140-6736(20)31927-9
125. ABRELPE. Associação Brasileira de Empresas de Limpeza Pública e Resíduos Especiais (2021) Abrelpe no Combate a Covid-19. Recomendações ABRELPE para a gestão de resíduos sólidos durante a pandemia de coronavírus (Covid-19). https://abrelpe.org.br/abrelpe-no-combate-a-covid-19/
126. Hanafi E, Siste K, Limawan AP, Sen LT, Christian H, Murtani BJ (2021) Alcohol-and cigarette-use related behaviors during quarantine and physical distancing amid COVID-19 in Indonesia. Front Psych 12:622917. https://doi.org/10.3389/fpsyt.2021.622917
127. Hashmi PM, Zahid M, Ali A, Naqi H, Pidani AS, Hashmi AP, Noordin S (2020) Change in the spectrum of orthopedic trauma: effects of COVID-19 pandemic in a developing nation during the upsurge; a cross-sectional study. Ann Med Surgery 60:504–508. https://doi.org/10.1016/j.amsu.2020.11.044
128. Instituto de Pesquisa Econômica Aplicada (2016) Desafios do desenvolvimento. Brasília: ano 12, nº 86. http://www.desafios.ipea.gov.br/
129. Instituto de Pesquisa Econômica Aplicada (2021) Indicadores Brasileiros para os Objetivos de Desenvolvimento Sustentável. https://odsbrasil.gov.br/
130. Lei nº 13.982, de 2 de abril de 2020. http://www.planalto.gov.br/ccivil_03/_ato2019-2022/2020/lei/l13982.htm

131. Medida Provisória nº 936, de 1º de abril de 2020. http://www.planalto.gov.br/ccivil_03/_ato
      2019-2022/2020/mpv/mpv936.htm
132. Qureshi AI, Huang W, Khan S, Lobanova I, Siddiq F, Gomez CR, Suri MFK (2020) Mandated
      societal lockdown and road traffic accidents. Accid Anal Prev 146:105747. https://doi.org/
      10.1016/j.aap.2020.105747
133. Szylovec A, Umbelino-Walker I, Cain BN, Ng HT, Flahault A, Rozanova L (2021) Brazil's
      actions and reactions in the fight against COVID-19 from January to March 2020. Int J Environ
      Res Public Health 18(2):555. https://doi.org/10.3390/ijerph18020555

COVID-19 and Sustainable Development Goal 12: What Are the Impacts of the Pandemic on Responsible Production and Consumption?

Ritanara Tayane Bianchet, Ana Paula Provin, Valeria Isabela Beattie, and José Baltazar Salgueirinho Osório de Andrade Guerra

Abstract The COVID-19 pandemic caused by the SARS-CoV-2 outbreak has resulted in a serious threat mainly to public health, but it is known that, in parallel, it affects all areas of sustainability: social, economic, and environmental. Consequently, achieving the Sustainable Development Goals (SDGs) becomes even more unattainable. In this context, the SDG 12 stands out specifically, which stood out in the face of the scenario of demand and continuous use of disposable materials, to avoid contamination and proliferation of the virus, directly impacting the generation of waste, which is commonly deposited in landfills and when improperly disposed of, they accumulate in improper places such as marine environments, sewage networks, or even open dumps. Also, steps to ensure cleaner and more sustainable production have become chimerical, given the lack of raw materials, the paralysis of large industries, and the increase in unemployment. It should be noted that, even if there is a decrease in the emission of greenhouse gases due to social isolation, it is understood that other areas have emerged and, this period, is not enough to mitigate the entire climate change scenario, therefore, even the "positive" impacts become unsustainable mainly in the long run. Given this situation, this chapter intends to contextualize some goals and indicators of SDG 12, such as goals 12.2, 12.3, 12.4, and 12.5, and consequently, analyze the impacts caused by the pandemic in the context of COVID-19 on responsible production and consumption in industries such as the health, food and beauty sectors, globally. For this purpose, high-impact scientific articles were used, researched through databases, official government documents, and the United Nations Development Program itself.

R. T. Bianchet (✉) · A. P. Provin
Environmental Science Master's Program, University of Southern Santa Catarina (Unisul), Av. Pedra Branca, 25, Palhoça 80137270, Brazil

V. I. Beattie
Elliott School of International Affairs, George Washington University, Washington, DC 20052, USA

J. B. S. O. de Andrade Guerra
Permanent Professor of Graduate Studies in Administration and Environmental Sciences, University of Southern Santa Catarina (Unisul), Av. Pedra Branca, 25, Palhoça 80137270, Brazil
e-mail: baltazar.guerra@unisul.br

Keywords COVID-19 · Sustainable development · SDG12 · Sustainable
production · Responsible consumption

1 Introduction

Registering more than 2 million deaths by mid-January 2021, COVID-19 demonstrates the severity of this coronavirus strain, which was first recorded in Wuhan, China, most likely because it found favorable conditions for its pandemic awakening. Due to climate change, deforestation, environmental changes, loss of biodiversity, and several other factors, the emergence of zoonoses and mutations of pathogens has become more frequent and favorable, many of them manifesting themselves dangerously and lethally for humans and animals [1, 2].

Coronaviruses usually have bats as essential mediators, after all, they host more zoonotic viruses than any other order of mammals. Researchers point out that bats are also a locus of coronavirus evolution, as the frequency of coronavirus recombination is high. Variants of viruses can therefore appear and be kept in bats with properties that greatly increase their aggressiveness. Two other coronaviruses had already caused, to a lesser extent, serious illnesses, such as SARS-CoV, responsible for the 2002–2003 SARS outbreak, and MERS, which caused an outbreak in the Middle East in 2012 [1, 3].

In the case of SARS-CoV-2, Malaysian pangolins were initially suggested as the hosts of the virus, since they found these sick animals in the illegal wildlife market in China, however, a subsequent phylogenetic analysis and amino acid sequence in the SARS-CoV-2 protein S, cast doubt on the hypothesis of pangolins as hosts. Therefore, they have not yet identified a real intermediate host, and the hypothesis that the virus has overflowed through bats to humans seems more acceptable and concrete [1].

Among the main reports of the worsening of people affected by SARS-CoV-2, there is acute lung injury (ALI) and acute respiratory distress syndrome (ARDS) until pulmonary insufficiency is achieved. The scientific community has been aware of the fact that the transmission of SARS-CoV-2 is greater than SARS-CoV due to the genetic recombination of the S protein in the RBD region of SARS-CoV-2, the contagion by COVID19 increased rapidly at the global level, therefore, on March 11, 2020, the World Health Organization (WHO) declared the emergence of a new pandemic [2, 4].

The entire world population has been aware of the signs and symptoms of the new virus. Among them are fever, cough, fatigue, shortness of breath, myalgia, sore throat and head, episodes of diarrhea and stomach pains, chest pains, and in some cases even skin blemishes. Symptoms can be mild to moderate with an incubation period of 6–41 days. There are also records of asymptomatic people. Therefore, because of the main forms of contagion and proliferation of the new virus, through exposure to coughs, sneezes, respiratory droplets, or aerosols, the worldwide orientation to mitigate this contagion was the use of masks for the entire population. Thus, the chances of contamination are lower, and consequently, hospitals fill up since other diseases continued to exist and affect people [4, 5].

In addition to the use of masks, the best way to ease contagion is non-agglomeration, that is, social isolation. With less human contact, there is less proliferation of the disease. So, society has to adapt to new norms of coexistence, with strict distancing, use of masks, and constant hygiene of hands and objects of common use. However, when it comes to hospital environments, the contagion is greater, since patients with symptoms or patients in a critical condition are treated, hospitalized, and under medical care. There may be cross-contamination between health professionals and patients, so the need for constant changes in personal protective equipment arises (PPE) and the need for disposable versions of this essential equipment also increases. [6].

Among PPE's, the most used are masks, gloves, caps, face protectors, goggles, lab coats, and shoe protectors. Therefore, the worldwide search and surge in demand for personal protective equipment began, and the high number of hospitalizations caused concern about the availability of these PPE's, registering one of the first obscurities of the pandemic. In view of this, emergency measures began to take place, such as washing and reusing items that should be for single-use, such as the case with N95 masks, which are the most efficient for aerosol filtering. Months after the current pandemic, there is still a lack of N95 masks in supply for health professionals [6, 7].

Despite all the issues involving PPE in hospitals, the most effective method for preventing mass contamination and the overload of the health system is social isolation except for essential excursions such as grocery shopping and picking up medicines. The first months of isolation shifted worldwide attention to health professionals and individuals in risk groups. The population remained alert and reclusive mainly because they did not know what would come of this new virus. Concerns included but were not limited to symptoms, consequences, and the release of a vaccine. However, over the weeks, with worldwide scientific commitment and engagement, some flexibilities started to emerge, such as the opening of some services with all possible protection measures, since the mechanism of action of the virus had already become more evident [8, 8].

Therefore, as we experience this pandemic and observe its various emergencies related to health and sanitation, other concerns begin to capture our attention, such as environmental issues and their impact (whether positive or negative). For [10] some, much of the pandemic of COVID-19, focused around the need to accelerate sustainable development since it correlates directly with the climate change emergency.

According to [11], the pandemic of COVID-19 has become a major public health issue worldwide, and consequently, it can impact environmental sustainability, social responsibility, and people's overall quality of life. In this context, environmental awareness, sustainable consumption, and social actions of people have been significantly altered with factors such as quarantine, social isolation, and the health crisis caused by the pandemic.

Therefore, it is necessary to discuss these consequences in line with the well-known proposals of the Sustainable Development Goals (SDGs), because progress must be made soon according to the 2030 Agenda. For [12], the pandemic and coronavirus disease originated in 2019 (COVID-19) which diverted government attention from long-term sustainability goals. Which in turn imposed unparalleled injections of resources to rescue national economies, with the risk of delaying the achievement of the Sustainable Development Goals (SDGs).

According to [13], it is worth remembering that the SDGs are a new universal set of targets and target indicators, which were adopted on September 25, 2015, to "eradicate poverty in all its forms" by 2030 "and balance the three dimensions of sustainable development: economic, social and environmental."The SDGs are made up of 17 objectives, which are subdivided into 169 goals, and each country is asked to incorporate these ambitious goals into its political agendas and to work to achieve the SDGs.

It is noteworthy that progress towards the goals before COVID-19 was already considered slow, and with the advent of the pandemic, brought immense additional barriers such as the contraction of economies and social inequalities, which are increasingly exacerbated. However, according to [10], it would be a mistake to conclude that the SDGs are somehow less relevant now, as the structure of the SDGs remains the best guide for identifying integrated solutions to protect the health of current and future generations simultaneously. At the same time, it manages to mitigate negative trade-offs during the COVID-19 crisis and the subsequent recovery.

Due to the large number of conflicts that must be researched and discussed in the face of this pandemic associated with the SDGs, this chapter seeks to cut the content and stick to the analysis and discussion related to objective 12 (Responsible consumption and production). The choice of SDG 12 was due to the increased generation of both domestic and hazardous biomedical waste [14, 15], as well as issues surrounding production and consumption by some economic sectors such as food, textiles, and packaging. Therefore, despite the notoriety of the positive environmental effects that resulted from less road traffic and cleaner aquatic ecosystems and atmosphere, solid waste management has expanded [16, 17].

For this, 4 of the 11 goals of SDG 12 were selected to carry out a more profound and improved search on the impacts caused by COVID-19 in view of this approach. The goals selected for the writing of the chapter "COVID-19 and the Sustainable Development Objective 12: what are the impacts of the pandemic that can be observed in responsible production and consumption?" were:

12.2: By 2030, achieve sustainable management and efficient use of natural resources;

12.3: By 2030, halve global food waste per capita, at retail and consumer levels, and reduce food losses along production and supply chains, including post-harvest losses;

12.4: By 2020, achieve environmentally sound handling of chemicals and all residues, throughout their entire life cycle, under agreed international milestones, and significantly reduce their release to air, water, and soil, to minimize their negative impacts on human health and the environment;

12.5: By 2030, substantially reduce the generation of waste through prevention, reduction, recycling, and reuse.

The choice of these goals, based on the authors' understanding, comprises key points for the debate on environmental issues, as it deals with the generation of waste from prevention to reuse, emphasizes the importance of efficient management, and brings two extremely urgent topics to be addressed: food waste and the environmentally sound handling of chemicals. Finally, the analysis of these goals and their indicators comprises the facts that occurred in the context of COVID-19, between the periods of the first reports on the virus in China until the completion of the writing of this chapter.

2 The Management and Sustainable Use of Natural Resources in the Context of COVID-19 (Target 12.2)

It has been known for many years that the changes in our climate and biodiversity, the scarcity of our principal resources, and the severe pollution associated with the impact of the use of resources, among others, is the result of the worldwide increase in population, material comfort, and technologies [18, 19]. It is noteworthy that the projection for 2050 shows that the world population may be 9.7 billion. According to [19], these growing and intensified pressures compromise the natural resource base and the ecosystem's absorptive capacity, affecting human health, the economies of nations, and sustainable progress.

Given unsustainable consumption and production patterns, following projections, cities will continue to emit increasing amounts of CO_2 equivalents, from heating and cooling technologies, transportation, production, and consumption of goods and services that are heavily dependent on fossil fuels. During the COVID-19 pandemic, substantial decreases were observed in some types of air pollution, but only time will tell whether these emissions will revert to pre-pandemic levels [18]. According to [20], in New York there was a 50% reduction in pollution due to the measures to contain the coronavirus, in China, CO_2 emissions fell by up to 25% and there was a 40% reduction in coal fuel from China's sixth largest power plants, according to reports published by the Ministry of Environment and Ecology.

Referring to the ancestral techniques of thousands of years of Chinese agricultural practices, the researchers [21], conducted a study with two indigenous ethnic groups in China (The Dong of Zhanli Village and the Buyi of Guntang Village in Guizhou Province). It was observed that indigenous knowledge plays an important role in the sustainable management of natural resources, but despite its importance and contribution to sustainable rural livelihoods and the management of natural resources, traditional knowledge and practice are rapidly disappearing in China.

And not only in China, but many local and indigenous communities around the world offer ecosystem services to the global public in general through the maintenance of natural resources due to their harmful use and management [22]. However, the lack of recognition of peoples' rights to land, unregulated and exploitative use of resources, and the unfair distribution of benefits resulting from the use of natural resources for private companies, were the common problems among all reported case studies by [22].

This continued disappearance of traditional knowledge as reported in China and elsewhere in the world, has suffered and is strongly influenced by rapid globalization and free mechanization under the "forced to develop" regime, including pressure from the "scientific development" discourse and "developing an innovative society and economy" [21].

It is important and necessary, first of all, to remember that good management and the use of natural resources maintain and/or increase the resilience of ecosystems and the benefits they provide. Consequently, following this reasoning, the possibility of implementing one of the most important concepts of sustainable development is favored and increased: "Sustainable development meets the needs of the present without compromising the satisfaction of the needs of future generations". In order for this thinking to become a reality, in addition to the Stockholm Conference in 1972 and the ECO-92 in 1992, the SDGs were created by the United Nations in 2015 and, among them, one of the goals covers protection issues of natural resources.

Goal 12.2 covers two issues of paramount importance to be discussed and which must be achieved by 2030: sustainable management and the efficient use of natural resources [23]. To this end, two indicators were also developed in order to assess progress against the target, after all, assessing the environmental impact due to the production and consumption of goods and services is a crucial step towards achieving the Sustainable Development Goals (SDGs). Thus, the indicators are: 12.2.1 Material footprint, material footprint per capita, and footprint material per GDP, and 12.2.2 Domestic material consumption, domestic material consumption per capita, and domestic material consumption per GDP [24].

Therefore, thinking about the path towards a sustainable future through green economies and the resilience of ecosystems, requires large investments and strategies. For [25], while the Sustainable Development Goals (SDGs) have attracted widespread interest and commitment from many organizations and stakeholders, current trends in terms of action to achieve the goals are not reassuring. With regard to the current moment, and with the crisis established by COVID-19, they make the search for global sustainability goals even more dependent on urgent investment decisions to be made by both public and private sectors [12].

To better exemplify this topic, "natural capital" will be subdivided for a deeper and more coherent analysis. Namely, "Capital" is a term used in economics to denote a stock of assets used to provide a flow of funds to establish or manage a business. Therefore, the term "natural capital" has been the subject of different interpretations, but the consensus view is that it includes the stock of biotic and abiotic assets in nature. Thus, the World Forum on Natural Capital defines "natural capital" as "the world's stocks of natural assets, which include geology, soil, air, water and all living things" [26].

2.1 Biological Resources

The misuse of natural resources through exploitation such as deforestation, droughts, and massive forest fires, has caused soil erosion in natural and environmentally protected areas rich in biodiversity, which in turn has caused these exploited areas to be the focus of international concern and research on the zoonotic origins of many global pandemics [18, 27, 28]. The interconnected causes and effects are resulting in increased food insecurity and an increased likelihood of future global pandemics like COVID-19, thus creating existential risks for global societies [18].

Examples in addition to the SARS-CoV-2 virus that caused the global pandemic of Covid-19, some of the other diseases transferred from animals to humans in recent years include Ebola, bird flu ('bird flu'), H1N1 flu ('swine flu'), Middle East respiratory syndrome (MERS), Rift Valley fever, sudden acute respiratory syndrome (SARS), West Nile virus and the Zika virus [27, 29].

In 2016, the United Nations Environment Program (UNEP) acknowledged the occurrence of a global increase in zoonotic epidemics, including 75% of emerging human infectious diseases, the origins of which were closely linked to environmental changes. These results are due to the fact that human beings are increasingly interacting and impacting ecosystems, given the close relationships between human, animal and environmental health [27, 30].

According to recent studies by [30], most pathogens move from their reservoirs of wildlife to human populations through hunting, consumption of wild species, and trade in wild animals along with other contacts with wildlife. It is also noteworthy that changes in the Earth's climate and weather continue to impact the planet's ecosystems, which include environmental communities with infectious disease agents and hosts that act as vectors.

In China, to reduce the risks in the human-wildlife interaction and interface, and to prevent possible epidemics in the future, a total ban on the consumption of terrestrial wildlife, whether artificially created or originally wild, and a ban on consumption of other wild animals required by the existing law in February 2020 [28]. According to [28], this change will affect some laws such as the Biosafety Law, Wildlife Protection Law, Animal Epidemic Prevention Law, and People's Law, among others, and the artificial commercial creation of terrestrial wildlife for food or food products and will promote wildlife conservation in China and other countries.

Reference [25] recall that in 2015, the Rockefeller Foundation-Lancet Commission introduced a new approach called Planetary Health and proposed a concept and strategy to safeguard human health in the time of the Anthropocene because human health and the health of the Planet should be one. Consequently, researchers, policy makers and regulators met at a conference in Helsinki in December 2019 to identify and discuss the main scientific challenges to improving and benefiting from healthy environments and respect for the integrity of natural systems [25].

According to the Convention on Biological Diversity (CBD), in its UN Global Biodiversity Outlook report (GBO-5) it was concluded that actions to protect biodiversity are essential to prevent future pandemics. In addition, the report notes that the loss of biodiversity can also lead to a faster rate of emergence and reemergence of infectious diseases [31].

The COVID-19 pandemic, which was identified shortly after the conference, exemplifies the extent of global interdependence and the lack of preparedness for threats to global health. The emergence of a new global pandemic was predicted, but the warnings were largely ignored. It was a question of when, rather than whether or how it would occur, considering how habitat disturbance, deforestation, live animal trade and intensive agriculture and land use increase the risk of zoonotic diseases in humans [18, 25].

Reference [32], recalls that the long-term effect of the synergy between the consumption of natural resources and environmental sustainability varies between countries, depending on the economic structure and the issue of environmental convergence between developed and developing countries remains inconclusive in the literature. Consequently, this situation emphasizes the need for a global partnership to achieve environmental sustainability.

According to studies by [32], using the ecological footprint instead of carbon dioxide emissions can provide a true and inclusive perspective on assessing environmental deterioration. For, the ecological footprint is responsible for built-up areas, carbon emission levels, cultivation areas, fishing areas, forest areas and pastures, according to data from the Global Footprint Network (GFN). Therefore, through the ecological footprint it is possible to capture all facets of the environmental dynamics. Through the Global Footprint Network (GFN,) website, it is possible to follow the data related to the Ecological Reserve and Deficit, the total Ecological Footprint by person and the total Biocapacity per person for any interested individual.

It is possible to observe through the data that most countries (>70%) are deficient in biocapacity, that is, countries in which the percentage of the ecological footprint exceeds biocapacity. Among the countries with a biocapacity deficit are the USA, India, China and most countries in Europe (marked in red). Among the countries that still have a percentage in which biocapacity exceeds the ecological footprint are Brazil, some African American countries and Australia (marked in green), to name a few examples. Finally, when it comes to the world, biocapacity per person is 1.6 gha, while the ecological footprint per person is in deficit of 2.8 gha, thus resulting in a global biocapacity deficit of -1.2 gha [33].

According to [18] it is notable that urban and regional planners need to work more more closely in adapting existing infrastructure and buildings in cities and planning new developments to incorporate more "blue" and "green" resources to reduce the "usual" challenges, but also to help them anticipate and mitigate the adverse effects of current and predicted pandemics. After all, reducing the richness and balance of biodiversity causes the disappearance of a fundamental part of the ecosystem that serves as a "buffer" for the spread of infectious diseases in humans, animals and plants. In this sense, several studies have suggested that the transmission of diseases increases with a loss of biodiversity [30].

2.2 Mineral Resources

According to [26], although human development depends on nature, non-living (abiotic) natural resources and processes are persistently neglected in international and national policies that promote sustainable development. And for society to continue to live sustainably on the planet, it is important that we understand the values that geodiversity brings and act to conserve, manage, and plan these geo-resources, which can be understood as "the natural extension (diversity) of geological characteristics (rocks, minerals, fossils), geomorphological (landforms, topography, physical processes), soil and hydrological characteristics [26].

Soil, for example, plays a key role in the conservation of biodiversity, as it provides essential services for agricultural production, plant growth, animal housing, biodiversity, carbon sequestration and environmental quality, which are crucial for complying with SDGs. However, soil degradation occurs in many places around the world due to factors such as soil pollution, erosion, salinization and acidification [34].

Mineral extraction practices have long been questioned due to poor management of natural resources and concerns about planetary limits. However, with the advent of the COVID-19 pandemic, borders were closed impacting the demand for material flow. According to [35], mining operations have suffered a dramatic impact due to the health and access of employees. Many countries with dominance in the mineral supply chain, such as China, Chile, and Peru, have faced a large number of viral cases and several mines have opted to stop operations to control the spread of the disease. Reductions in metal prices have reported that the mining industry will be adversely affected on the demand side by the COVID-19 crisis in the short term. However, the low price elasticity prevented a crisis of scarcity in the supply of mineral resources [35].

According to the researchers [35], there is a global consensus to guarantee the openness, stability and security of the global supply chain of mineral resources and, among the challenges, are: (1) Artisanal and small-scale mining (ASM) activities will be seriously affected, (2) The global distribution of the metal supply chain would be reorganized; (3) Mining technology can find bottlenecks and possible areas for growth in a post-pandemic world include metals to increase digital infrastructure; and (4) The increase in the cost of transport in international trade in mineral resources.

The activities of artisanal and small-scale mining (ASM)—extraction and processing of low-tech and labor-intensive minerals—in Sub-Saharan Africa, suffer a great impact due to COVID-19. According to the researchers [36], scholars have shown evidence that ASM is the most important non-agricultural rural activity in sub-Saharan Africa, as it provides a complementary source of income and the sector helps millions of poor and farm dependent families in the region deal with serious economic problems.

It is noteworthy that Sub-Saharan Africa has managed to prevent a high number of infections and deaths from COVID-19 at the time of writing this article, but has suffered from the economic impacts of the pandemic. According to [36], perhaps no other place has suffered more than sub-Saharan Africa in its remote rural areas, which are already affected by poverty and produce food mainly at subsistence levels.

The pandemic shows the need for well-designed and efficient management strategies. To this end, the United Nations Environment Assembly passed a resolution in 2019 requesting member states to consider coordination on the governance of mineral resources, following the report of the International Resource Panel.

Thus, some processes were considered: (1) The quantitative and qualitative data collected during the pandemic need to be consolidated to identify the main vulnerabilities in the supply chains of minerals, both from mined and recycled sources; (2) The establishment of a stock system for items such as oil to reduce the impact of emergencies on the supply of mineral resources; and (3) Given the growing concerns about the social and environmental performance of industries, which frequently suffer during crises, a mechanism should be put in place to ensure the rationalization of methods for assessing comparative impacts [35].

2.3   Water Resources

The issues involving water resources and COVID-19 stood out, mainly, in relation to the likelihood of water contamination and consumption during isolation through the virus. All of these topics hinder the management and the proper functioning of this resource to meet the demand of diverse populations, as well as, sustainability.

In 2003, in Hong Kong, a SARS outbreak involving 321 patients was attributed to the contaminated sewer system. Recently, researchers reported that live SARS-CoV-2 can be isolated from the excrement and urine of patients with COVID-19, and the viral nucleic acid of more than 20% of patients remained positive in the stool samples even after viral elimination in the respiratory tract (Wang et al., 2020). According to [37] and [29] the presence of COVID-19 in faeces has been significantly reported in the literature and the presence of faeces in sewage drains leading to groundwater contamination may be an emerging threat to water pollution and can lead to the spread of COVID-19.

Thus, according to [29, 37] and Wang et al. (2020), in relation to the risk of water transmission of SARS-CoV-2, studies should not be neglected to answer whether hydrological conditions could be associated with the outbreak of COVID -19. It is known that emerging pathogens can enter wastewater systems by eliminating pathogens in human waste, introducing decontamination wastewater, illicit activities, breeding animals and hospital effluents, or draining surface water after a biological incident [29].

It is noteworthy that waste water is all water coming from urban public residences and facilities (hospitals, schools, etc.), as well as from certain industries (if it does not require specific treatment). This water is sent, through the "sewage network", to treatment plants, where it is treated and then released into the environment [29].

According to [29], it is very difficult to track the virus because most people are asymptomatic and, in addition, it is not possible to perform active clinical tests on all individuals, due to resource and cost restrictions. In addition, COVID-19 can also display two or more waves. In these circumstances, the passive but effective method of monitoring sewage or effluent can be used to track and trace the presence of SARS-CoV-2, through its RNA of genetic material, and examination of the entire community.

When it comes to consumption, a study was carried out with 27 countries in the European Union and it was observed that only 11 of these countries (40.7%) implemented at least one policy intervention that considered the water sector. It is recalled that water supply and sanitation have become central, since washing hands under running water is the main approach to mitigate the spread of COVID-19. However, maintaining hygiene practices and mitigating the COVID-19 pandemic can be challenging, especially in places where freshwater sources are scarce [38].

For [38] and collaborators, there is still growing uncertainty about how water availability and management in various countries can face the challenges of the COVID-19 pandemic, since COVID-19 changed the patterns of water consumption and also made more acute impacts of drought and perennial water scarcity in countries such as Ireland, the United Kingdom, Turkey, Ethiopia, Kenya, Syria, Poland, Romania, Kosovo and India. These countries already suffer from climatic variation and continuous decline in precipitation patterns.

Ireland's national water company reported that there was a 20% increase in general residential water demand during the blocking period. In Baden-Wurttemberg, Germany, the municipal water utility Stadtwerke Karlsruhe (SWKA) also revealed variations in water consumption, which were attributed to the closure of schools and non-essential activities, and the increase in domestic work. Records from Portsmouth, England, also showed a 15% increase in water demand among consumers and a 17% reduction for non-domestic consumers during the peak period of the pandemic [38].

Researchers in Brazil, specifically in the city of Joinville, located in the south of the country, carried out an assessment of the impact of actions to prevent the spread of the coronavirus (COVID-19) on water consumption. The analyzes were applied to the commercial, industrial and public consumption categories and Wilcoxon and Kruskal–Wallis non-parametric tests were applied and the Prais–Winsten regression models were adjusted.

According to [39] and collaborators, the results of the Wilcoxon test showed that there are differences between the periods analyzed, indicating a drop in water consumption in commerce and industry, but there was an increase in residential consumption. As for the results of the regression model, they confirmed the effect of restrictive actions in reducing consumption in non-residential categories. The results also indicate an increase in water consumption, which was more pronounced in apartment buildings than in houses, isolated or grouped in condominiums.

It is noticeable that the pandemic has affected countries with different cultures, economies, and policies. This chapter discusses some situations and studies within a wide range of surveys where they show and report on countries in which there is often no basic sanitation and piped water for the majority of the population. Therefore, it is observed that understanding the aspects related to water consumption is important to gather essential information to ensure that the urban water supply system is resilient in a pandemic context. Mainly, with regard to the search for sustainable development, we perceive many issues to be managed and solved, both for the sake of sustainability and for the well-being of people, especially in epidemic situations.

2.4 Energy Resources

Energy resources are vital to the economic development of any nation and play an essential role in human life. Energy use in most countries increases over time due to social changes, lifestyle, population growth, architectural and urban projects that are not environmentally friendly. In 2018 alone, the European Union noted that world energy consumption reached 9,938 Mt of oil equivalent (Mtoe) and this growth was mainly driven by China, the United States and India, which together account for about two-thirds of growth [40].

Many countries are currently experiencing varying levels of energy crises due to limited natural resources, along with the Covid-19 pandemic. This instability can lead to the shutdown or restriction of many industrial units, limited access to energy and worsening unemployment. It should be noted that the main reason for these problems is the widening gap between energy supply and demand, logistics, financial issues, as well as ineffective strategic planning issues [40, 41].

According to [41], statistical and projection data from the International Energy Agency (IEA), state that the shock in energy demand in 2020 should be the biggest in the last 70 years. And, according to the research by [40], the International Energy Agency mentioned that the initial goal is to reduce annual energy by 2.6% by 2030. However, the rate of increase in global energy intensity was only 1.7% in 2017, and an increase to 2.7%/year is necessary until 2030. In 2018, an increase in global energy intensity was only 1.2%. This means that from 2019 to 2030, the global energy intensity must increase again by 2.9% each year to achieve the goals and objectives of sustainable development.

If we analyze the blocking period (either partial or total) and with the reduction of many economic activities, the global energy demand has consequently decreased [41, 42]. The latest data from July 2020 show that, compared to the same period in 2019, peak rates of reduced electricity consumption (corrected by the climate) in France, Germany, Italy, Spain, the United Kingdom, China and India during the blocking period were more than 10%. However, the drop in demand and energy consumption is damaging to the energy sector, as is the case in the United States, which led to the bankruptcy of at least 19 energy companies in the industry during the pandemic [41].

Among the established losses, the global crude oil market experienced a significant fall after the new coronavirus outbreak (COVID-19). All major oil markets have become extremely volatile investments in these markets can lead to substantial losses [20, 43]. Demand for oil has witnessed a dramatic decline in industrial, commercial, domestic and transportation uses. This led to enormous pressure on the balance between supply and demand and storage to control oil prices (Siddique et al., 2021).

The fall in oil prices in the midst of the pandemic led to greater risk probabilities for oil-derived assets, in general, oil prices fell by 40% (Siddique et al., 2021). In the Middle East and North Africa, for example, there is a sudden drop in the price of oil. In the world superpower of the American economy, the reduction in the price of oil is impactful, the S&P 500 predicted that the American economy would face the worst situation on Wall Street since the financial crisis of 1929 [20]. It is also noteworthy that the Organization of Petroleum Exporting Countries (OPEC), represented by Saudi Arabia, interrupted negotiations with Russia on further cuts in oil production, oil prices plummeted and futures showed negative oil prices in March [35].

The future of oil will largely depend on how countries respond to the COVID-19 outbreak and how they will recover their economic activities. According to Siddique et al. (2021), this crisis would result in a major cut or delay in other ongoing or future development projects in hospitals, roads, hotels, tourism, industries, energy production, education and many others. A prolonged period of low oil prices would not only reduce conventional energy resources, but could also bring about a paradigm shift in renewable resources. This could lead to a shift in economic dependence on oil to other alternative commodities.

In contrast to falls and industrial crises, on a macro scale, the increase in residential energy consumption must be considered comprehensively for a better conclusion on energy demand [41, 42]. It was observed that the demand for electricity fell for commercial and industrial loads, but increased for residential loads for most of the countries studied, as industrial and commercial activities were restricted while people imposed permanence and work at home (Elavarasan et al., 2020).

A survey carried out in California, during the social practice of active isolation (May 5–18, 2020), showed some changes in domestic occupation patterns. In addition to the work done at home, an increase in occupancy at noon (10 am–3 pm) was reported, mainly by young people and minors and, consequently, an increase in the rate of energy use. However, it was also a trigger to rethink smart technologies for the home, as well as the origin of these energies [44].

The reduction in electricity demand has had an acute impact on the generation and supply of fossil fuels instead of cheaper renewables, in fact, renewables seem to accelerate the trend of fossil substitution [41]. Reference [42] highlight the position of the International Renewable Energy Agency (IRENA) and the Council of European Energy Regulators (CEER), which discussed issues such as the importance of fuel diversification, integration and sustainability of energy production, a positive effect on climate goals, conversion of the energy system into a "clean energy system", solar and wind energy becoming competitive, investment opportunities in renewable energy, and a switch to clean hydrogen.

A report published by the Union of the Electrical Industry—Eurelectric aisbl provided some recommendations on electrification and sustainability due to the impacts of COVID-19. The report in the "microgrid knowledge" stated that the construction of microgrids would have a progressive impact on the reconstruction of the economy and also in facing the challenges of the electricity and energy sector. According to the information, the development of a microgrid-based energy system facilitates intelligent controls, clean energy integration and the ability to support the grid during any crisis [42].

Therefore, it is important to have a strategic management process for global energy and to determine where the global energy transition should go. In doing so, policymakers must have a long-term plan that involves decision-making processes at all levels, including global, regional, national, state, municipal, district and sectoral. Although the nations have different visions, missions and energy management strategies according to their benefits, it is necessary to have an integrative management at a global level of all these countries, so that the production and consumption of all sources of energy are sustainable and meet all the needs of the community [40].

3   Food, Waste, Production and Supply in the Context of the COVID-19 Pandemic (Target 12.3)

The impact of COVID-19 on the food and agriculture sector exposed vulnerabilities in the agrifood supply chain, however, it should be noted that global food and nutrition security has been threatened since well before the arrival of the pandemic. More than 800 million people sleep hungry every night and more than two billion of the world population suffers from hidden hunger due to the lack of essential micronutrients [18, 27, 32, 68]. Therefore, sustainable food systems are needed to guarantee sustainable food production that meets the growing demand of the projected 9.8 billion people by 2050 by 2050 [45].

Overall, the short-term impacts of COVID-19 required immediate responses to limit the spread of infection through the implementation of health care and containment measures, as the agri-food system can be considered a public health instrument (Fan et al., 2021) [68]. There is no evidence that food is a likely route of transmission of SARS-CoV-2, however, the potential risks of infection through food and water cannot be ignored, such as the high persistence of the virus in food and surfaces (FAO 2020) [46].

Meat from cattle, poultry, pigs and wild animals, for example, are known to be abundant in heparin sulfate, which is necessary for SARS-CoV-2 to interact with the host tissue epithelium. The persistence of this virus in the environment and on food contact surfaces, such as plastic, wood, rubber and stainless steel, means that it can survive for several days, so meat tissue surfaces can be a potential route of transmission or even critical for COVID-19 infection [46].

Reference [45] hope that food security challenges related to the pandemic will slow progress towards eliminating hunger and poverty in sub-Saharan Africa. For example, 55% of the world's hungry people and 70% of the world's poorest people lived in Africa before COVID-19. Thus, the pandemic has aggravated an existing inefficiency in the production, distribution and consumption of food in the region, canceling the advances in the eradication of hunger and threatening to reverse the gains obtained in the fulfillment of the Sustainable Development Goals (SDGs).

According to [47], there is wide media coverage of sudden drops in food security due to some factors such as: (1) Loss of income of workers totally or partially dismissed, (2) Requests for stay at home and restricted physical access to food markets and/or indigenous food collection activities; (3) Closing or decreasing the capacity of institutions that support social food safety nets; (4) Market disturbances, such as problems with the supermarkets' ability to replenish quickly and waste of fresh vegetables; (5) Fruit and milk due to the inability of farmers or entrepreneurs to transport them.

According to [48] and [45], the agrifood supply chain connects producers to consumers, incorporating activities in the farm, post-harvest processing and manu-facturing, commerce and distribution, retail, service sector supply and regula-tory processes to guarantee quality and safety. Thus, if the COVID-19 pandemic continues, the production of agricultural crops, aquatic foods and livestock such as wheat, rice, vegetables, fish, seafood, eggs, meat and dairy products will be greatly affected [50].

In Asia, despite the Association of Southeast Asian Nations (ASEAN) being the source of many agri food products for the world, data from the Asian Development Bank show that agriculture's contribution to national GDP has been declining since the 1990s and now ranges from a minimum (Singapore) to a maximum of 23.5% for Cambodia. At the same time, employment in the agricultural sector has also declined, ranging from less than 1% in Singapore to 33% in Cambodia [48].

According to the studies carried out by [49] in the Pacific Island region, it was found that the identified impacts include: (1) Reduction in agricultural production, availability of food and income due to the decline in local markets and loss of access to international markets, (2) Increase in social conflicts, such as land disputes, theft of crops and high-value cattle and environmental degradation resulting from urban–rural migration; (3) Reducing the availability of seedlings, planting materials, equipment and labor in urban areas; (4) Reinforcement of traditional food systems and local food production; (5) Re-emergence of networks and cultural security values, such as exchange systems.

The authors also point out that before COVID-19 there was a diversity of food that is being diminished, including the contraction of family income. However, it was observed that households in rural and urban communities in the Pacific Island region appear to have responded positively to COVID-19, increasing food production from home gardens, particularly tubers, vegetables and fruits [49].

Bangladesh, which is considered a low-income riverside country in Southeast Asia dependent on agriculture, which is divided into three sub-sectors called farming, fishing and livestock. Bangladesh is currently struggling to combat the associated adverse impact of COVID-19, as the virus is damaging the economy by disrupting industries, poultry, dairy, agricultural and aquatic food systems that also threaten the livelihoods of directly or indirectly dependent communities. indirectly. Stressing that the Food and Agriculture Organization (FAO) reported that people who are already malnourished, weak and vulnerable to disease are more likely and a "crisis within a crisis", combining the current health crisis with a hunger crisis [50].

Sunny et al. [50] report in their research that fish play a vital role in the diet of the people of Bangladesh due to the abundance of open waters inland (3rd place in fisheries for capture fish), closed waters inland (5th place in world aquaculture production) and marine fishing. The challenge, however, is to ensure uninterrupted production and supply of aquatic food, fair prices for products and also nutritious food for everyone.

The authors also point out that although Covid-19 did not directly interrupt the production of aquatic food, it hinders the complexity of transportation and the weak presence of the buyer causes an abnormal drop in the prices of fishery products. So it wasn't just fishermen, fish farmers, retailers, entire sellers and other members of the aquatic value chain that have been affected, but also the country's general economy [50].

The massive reduction in demand for restaurants and commercial food services in combination with restrictions on labor, processing capacity and/or storage has led farmers to discard their mass production. Quarantine measures are seriously affecting the availability of labor for time-critical agriculture, from planting vegetables to harvesting fruit. As the crisis develops, these impacts are likely to become more widely and deeply felt in agricultural sectors and national economies [47, 51].

It should be noted that the pandemic substantially changed normal living conditions and created a "new normal" that forced economic and socio-behavioral changes in the absence of medical treatments and vaccines. In the search for a new normality in people's way of life, there is an increasing tendency to avoid the consumption of dinner services, which involve physical contact with people [52].

During the first wave of the pandemic in mid-March 2020, most restaurants were forced to suspend dinner services, and only take-away, drive-thru or delivery services were allowed. The National Restaurant Association survey (2020) showed that the restaurant industry lost more than $120 billion in sales and 8 million employees were laid off or laid off in May 2020. It was predicted that the pandemic could cause losses of up to $240 billion by the end of the year. Although restaurants in all states could later reopen, restaurant operators still required essential health information to minimize the risk of spreading COVID-19 [53].

In the USA, for example, COVID-19 and subsequent restrictions on restaurant meals have negatively affected the restaurant industry. Although restrictions on meals have been adopted to avoid human contact, the evidence suggests that consumers may erroneously perceive that the restaurant's "food" and its "packaging" are risky sources of COVID-19. According to the research carried out by [54], consumers were less concerned with the contamination of COVID-19 from food in general than with restaurant food and its packaging, and Consumer perceptions of risk varied with financial concern for food, gender and being in a high-risk category of COVID-19 [54].

In the case of strategies to contain the crisis in the restaurant industry, a survey was carried out with a total of 86,507 small and medium-sized restaurants with sales data collected in nine cities in Mainland China. It was found in this study that the delivery service contributed to the sales of restaurants during the COVID-19 pandemic, casual restaurants and fast casual dinners had more benefits with the delivery service than fine restaurants, the effects of the brand in the financial performance of restaurants were stronger after the outbreak of COVID-19, however, discount options failed to increase sales during the pandemic [52].

More than half of the total amount of food wasted in Europe concerns domestic food waste, mainly due to incorrect food management habits and behaviors [55]. Changes in eating habits, as a consequence of lifestyle disruptions and psychological stress due to blockages, can produce an important point that can influence the patterns of generation and distribution of Food Loss and Waste (FLW) throughout the supply chain [55, 56]. Reports from the Ministry of Agriculture, Fisheries and Food (MAPA) of Spain show, in general terms, that household consumption increased significantly in all food categories during the pandemic [56].

Spanish consumers are stocking up on non-perishable food and other supplies, eating more indulgent and comforting foods, drinking more wine, beer and other spirits, as well as snacks throughout the day. Obviously, these behavior patterns imply not only changes in the food supply chains and the generation of FLW, but also repercussions in the food pattern, which can be harmful to health and also to other environmental attributes offered by the Spanish Mediterranean diet, triggering obesity, sleep disruptions or impacts on the immune system [56].

However, it is possible to observe some positive aspects in some locations during the COVID-19 pandemic. A study carried out in Italy investigated how these dramatic changes in the daily lives of consumers influence the generation of food waste at the domestic level. A questionnaire was administered to a sample of 1,078 Italian consumers during the blockade (March–April 2020). Respondents were asked to self-estimate the percentage of food that their families wasted before and during the block and to explain their eating habits [55].

The survey revealed that most families threw out less food during the COVID-19 blockade compared to the situation prior to COVID. The results also showed that young consumers and people who started to implement good food management practices (shopping list, meal planning, etc.) considerably reduced the food they wasted during the blockade. In addition, the logistical difficulties in purchasing food experienced by consumers during the blockade made them manage their domestic food consumption more carefully [55].

Finally, it is important to capture the immediate effects of the COVID-19 pandemic on agricultural and food systems in its broadest sense. As observed through current scientific research, the pandemic has weakened the food chain in different locations and in different sectors. Therefore, in order to achieve the Sustainable Development Goals, not only with regard to this topic on target 12.3, but including questions about world hunger and malnutrition, it is necessary to create strategies with a view to sustainable and consistent agriculture. the reality of more than 7 billion people in the world.

To this end, the Food and Agriculture Organization of the United Nations created a "FAO COVID-19 Responses and Recovery" program. According to FAO, unless we take immediate action, we risk a global food emergency that could have long-term impacts on hundreds of millions of children and adults. Accordingly, FAO is asking for $ 1.3 billion in initial investments to provide an agile and coordinated global response to ensure nutritious food for everyone during and after the pandemic. The COVID-19 Response and Recovery Program allows donors to take advantage of the Organization's calling power, real-time data, early warning systems and technical knowledge to target support where and when it is most needed.

More information is available on the Food and Agriculture Organization (FAO) website (http://www.fao.org/partnerships/resource-partners/covid-19/en).

4   Waste from the Pandemic and Its Impacts on the Achievement of Goal 12.4

The SARS-CoV-2 pandemic has certainly hit everyday life worldwide in a historic way. As of January 25, 2021, Covid-19 had already infected more than 99 million people, and of these, more than 2.1 million died [7], New York Times, 2021). There are predictions that this pandemic will not end so quickly, since the applications of the first doses of vaccines have recently started and, until all are properly immunized, there is a long time to go, consequently, the demand for PPE will not decrease and also, care of the population regarding the use of masks and social distance should not end as quickly [7].

When talking about an exorbitant increase in the production and consumption of medical supplies, we have as an example Wuhan in China, which consumed more than 240 tons of medical waste every day during the period of the outbreak, being about 190 tons more than in normal days. In India, the city of Ahmedabad doubled the amount of medical waste generated, increasing from 550 kg/day to close to a ton/day in the first weeks of the lockdown. Cities like Manila, Hanoi and Bangkok experienced similar increases, producing 150–280 million tons more medical waste per day than before the pandemic. However, this consumption goes beyond hospitals, the domestic demand for PPE is also significant [57].

The scientific community has been raising discussions regarding the reuse of materials that should be for single use, coupled with this, point to the need for the development of technologies that assist in the sterilization of this equipment in a safe manner, also raising questions regarding the preparation of PPE's with biodegradable materials, since they are mostly produced with petroleum products, finally, they have been analyzing the environmental impact of these materials. Data already point out that the global mismanagement of PPE contributed to environmental pollution with a monthly average of 129 billion masks and 65 billion gloves (Prata et al. 2020). Therefore, the aquatic environment is likely to be subject to more plastic pollution from these emerging debris items.

A study by [58] demonstrated that, among 660 health professionals from public and private hospitals, 424 (69.5%) had heard about the health waste disposal system but only 71 (11.6%) professionals had participated in training in a health waste disposal system, lighting a warning sign, since they are the main ones to deal with possibly contaminated materials and they must have a correct destination. Regarding the availability of materials for the disposal of health waste in the institutions visited, the study showed that 211 workers answered that "always", 339 workers answered that "most of the time", 14 workers answered "never" and 96 professionals answered "rarely" [58]. It is exceptional that all hospital centers have the appropriate materials available for professionals to dispose of hospital waste, given the importance of avoiding the loss of potentially contaminated equipment.

In addition to health professionals, the population in general should have access to training in the correct destination of personal protection residues, especially in this pandemic scenario, avoiding potential sources of contamination and proliferation of other diseases in addition to COVID-19. For example, in Toronto, Canada, the city's waste management division issued Special Waste Disposal Instructions to inform residents about the disposal of personal hygiene and sanitary waste during Covid-19. These instructions suggest the disposal of PPE in garbage bags along with household waste that is destined for municipal waste management facilities. The implementation of an effective collection system with proper disposal is necessary to reduce the release of contaminated plastic from COVID-19 into the environment [59].

In Brazil, among the recommendations passed on by specialists to the population, there are the use of two garbage bags, followed by the identification of this bag— warning that they are masks or the possibility of it being contaminated—and finally, if possible, advise placing it together with the bathroom trash. Also, there are guidelines for when the person is out of the house, the used masks should not be placed in the street bins, as they leave potentially contaminated material exposed to solid waste collectors. The recommendation is to store them in a plastic bag and put them in the bathroom trash when they get home [60].

A survey by Ammendolia and collaborators [59] sought to quantify and identify the densities of waste from personal protective equipment in the city of Toronto in Canada. There were 5 weeks of collections and analyzes. The researchers used a mobile application made just to document debris in any environment, then recorded the PPE residues they found at ground level and the researcher's height. They recorded a total of 1306 items in a cumulative area of 245,190 m^2 in six sampled locations. Of these items, 571 were gloves, of which 337 were identified as nitrile, followed by vinyl ($n = 154$), polyethylene ($n = 41$) and latex ($n = 39$). Most of these gloves were found close to a large market, which was associated with the use of gloves by the population to make purchases, after all, the use of them presents a sense of security, when the ideal is to properly wash your hands and to guarantee the destination of these gloves for the front line professionals [59].

Still in this study by [59], the masks they found were mostly surgical masks ($n = 381$), followed by reusable masks ($n = 12$) and dust masks ($n = 8$). This finding corroborates the fact that they find a large part of these masks near a hospital, in which visitors receive disposable masks to access, and consequently, do not dispose of them correctly, throwing this debris through the streets [59]. In view of this, we note the need for national and international policies that direct the population to the correct disposal of this material.

In addition to citizens, trade began to adapt to the pandemic scenario. As the flexibilization of distance became apparent, new rules began to be established, for example, in public environments, the use of masks, hand sanitizers, decreased maximum capacity of regulars began to be required, and also, single-use plastics began to be used. visas. Bars, restaurants, cafeterias, clinics, stores and other departments, started to insert protective plastics in cardboard machines, chairs, tables, cutlery, gloves to serve in buffets, counters, in toilets, in order to be able to replace them when necessary, thus avoiding possible cross-contamination [15]. Single use, quick disposal, plus an environmental problem and overload in waste management.

In view of all the issues in debate about waste, SDG 12 brings some important goals to be debated even more in this pandemic scenario. Goal 12.4 also provides for "significantly reducing the release of waste into the air, water and soil, to minimize its negative impacts on human health and the environment". In view of this, Fig. 1

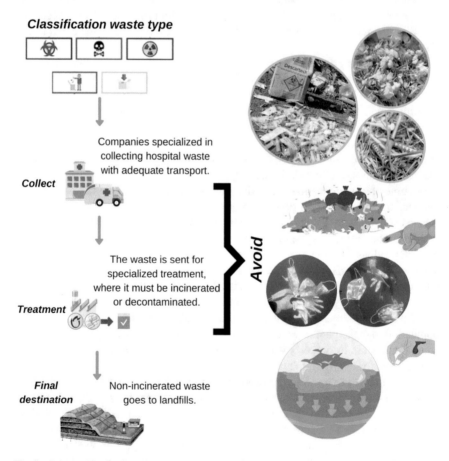

Fig. 1 Correct steps for hospital waste disposal. *Source* Authors, 2021

depicts the correct steps to be followed for a safe disposal of hospital waste, and secondly, how this correct disposal avoids greater environmental impacts.

First, materials must be discarded as to their type and risk. In the hospital, there is waste with physical, chemical and biological risks. From this separation, a specialized company collects this material and safely transports it to a treatment site using technologies such as microwaves, autoclave and incineration. When these possibilities are not available, the waste, after being treated, is deposited in landfills together with ordinary garbage. In this way, the disposition becomes safer, both for the environment and for human health.

In the face of inadequate management of these hazardous wastes, when there is contact or exposure to them, infections, dermatitis, cuts, neurological disorders, fever, hepatitis, AIDS and other viral contamination may occur through blood sharps. Generally, the people who are most affected by the incorrect disposal, when inside hospitals are the health professionals themselves, patients and patients visitors, during the collection process, are the workers of waste disposal services, finally, collectors and people working in landfills or open dumps, and citizens at greater vulnerability [58].

After improperly deposited, dangerous substances present in hospital waste, such as medicines, or the very decomposition of these materials together with common or organic waste, can affect the soil and water, especially the water tables, compromising the natural resources used throughout population and bringing serious consequences to the environment [58]. In addition to the PPE that is mostly produced with plastic derivatives, normal consumption of routine plastics still remained, data from 2018 suggest a generation of plastic waste of 6.9 megatons, of which 42% were deposited inefficiently, or that is, thrown in any trash or even on the ground, rivers, culverts, or are placed in a sanitary landfill together with common and organic waste. Thus, poor waste management threatens the ability of the global community to meet the goals of the Sustainable Development Goals [15].

Pollution of air, soil and followed by groundwater pollution due to the decomposition of waste has been the focus of research for a long time, among the main harms is the formation of leachate that if not treated it can cause greater risks of environmental contamination, it is also a source of organic acids that are converted into CH_4 and CO_2, requiring strict control over the generation of leachate from these residues. Another problem is related to gases originating from decomposition, such as methane and carbon dioxide, both major contributors to the greenhouse effect and, consequently, climate change [61].

Figure 2 clearly demonstrates how the rampant consumption of plastics and their derivatives, especially abruptly, already impacts the entire terrestrial and aquatic ecosystem, and the trends and consequences expected in the face of the pandemic.

The figure shows the presence of microplastic in the soil, from garbage and decomposed bags, as shown by some samples collected from landfill residues, in which the main types of microplastics found are derived from polyethylene (PE), polypropylene (PP), polystyrene (PS), expanded polystyrene (EPM) and urethane polyether (PEUR). PE plastic is the most used worldwide, usually in the production of packaging and personal care products. In addition, polypropylene (PP), the second most

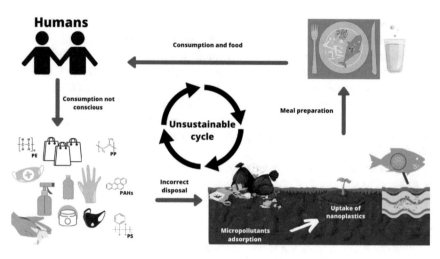

Fig. 2 Unsustainable cycle of plastic discarded incorrectly. *Source* Authors, 2021

common type of plastic found, is the most used in the production of disposable face masks, widely used in the pandemic scenario [62].

In the soil, microplastics affect all fertility, since they affect animals that play a fundamental role in soil diversification, such as earthworms, ants and other organisms, which ingesting materials with low organic composition (e.g., microplastics), some animals of the soil will produce more intestinal mucus, thereby altering the soil microbiome. In addition, plants also grow with nanoplastics already infiltrated in their development [63].

When these microplastics (PM) reach river paths and flow into marine environments, then it becomes more difficult to control or attempt to mitigate, therefore, it is estimated that at least 5.25 trillion plastic particles weighing together almost 269 thousand tons are prevailing in the world's oceans. Given its microstructure, it is confused with food, being consumed by several marine biota including phytoplankton, zooplankton, corals, sea urchins, turtles and fish. What becomes worrying, because in the presence of MP the heterotrophic plankton performs phagocytosis, which retains these micro plastic fragments in their tissues, resulting in decreased chlorophyll absorption and impacting the entire aquatic food chain, as these organisms are important primary consumers [64]. In addition, countless marine animals have already been found dead by suffocation, as they are unable to feed themselves due to the presence of numerous micro and macro plastic particles in their stomachs.

Finally, the entire normal cycle of the ecosystem went through and the microplastic remained throughout all the processes, be they whole, or each time as more nanoparticles, but always with an end many times already established: on our plate, our glass of water, our stomach, our digestive process. Human beings, the same exacerbated consumers, are severely impacted, as the consumption of PM with a size less than 130 μm can release toxins, additives and monomers, triggering, consequently, a carcinogenic behavior. In addition, human ingestion of microplastics can cause damage to the lungs and changes in liver function. Thus, microplastic pollution must be understood and controlled to preserve human, animal and aquatic health [62].

Another alarming question is related to the incorrect disposal of medicines, since they have chemical components that are decomposed in inappropriate places, they can generate innumerable harm, mainly with regard to microbial resistance. Studies report that if the decomposition of a drug is directed to water or infiltrates from the soil to the water tables, it induces changes in the microbial population, and represents a potential threat to the aquatic ecological process. In addition to many antibiotics losing effectiveness and microorganisms gaining resistance, altering an entire biological system [65].

4.1 Generation of Excessive Packaging from Online Purchases

Social detachment has become the new reality for millions of people worldwide, but it was not a sign of a decrease in consumption. Online shopping was the highlight especially in the first months of isolation. Data from April 2020, suggest that in the same month e-commerce grew 81%, associating more time at home and in front of technologies such as notebook and cell phone, and even to avoid running the risk of infection, people have resorted to the online service. Purchases of beverages, food, personal care and electronics have grown significantly. As a result, it is known that for every delivery, there is an excessive amount of packaging behind.

Plastic bags, paper, cardboard box, bubble wrap, stickers, Styrofoam trays, and Styrofoam balls. These are just some of the main means used for online shopping deliveries. Parallel to this, the demand and disposal of these materials increases, becoming more waste generated on a larger scale during the pandemic. For example, it is necessary to develop new materials for delivery packages, with internal policies for reuse or accumulation of points to the correct destination of each package.

5 Recycling, Reuse and the Development of Biopolymers as Aids in Meeting Goal 12.5

Goal 12.5 predicts by 2030, to substantially reduce the generation of waste through prevention, reduction, recycling, and reuse. Therefore, there is a demand for knowledge in the stages of the production processes, materials used, useful life, product life cycle, which correct recycling process, efficient and safe ways to reuse. According to [66] waste recycling has always been a major environmental problem of interest to all countries, being an effective way to prevent pollution, save energy, generate jobs and income and, in addition, conserve natural resources. However, with the pandemic and the high risk of contagion, some countries interrupted their recycling

Fig. 3 Types of plastics and multiple applications. *Source* Authors, 2021

programs and closed material collection centers, other countries forbade infected residents to separate their waste, for example, we have Italy [66]. Consequently, what was already difficult to maintain, with the pandemic the situation worsened, making this goal more unattainable.

As already mentioned, there are several types of plastics for multiple applications, as shown in Fig. 3.

Each type has a code with an identification number as shown in Fig. 3. Therefore, it is important to check with the company that does the recycling, or with the local city hall, which types of plastic are accepted. Ideally, the plastics are clean upon delivery, dry, and divided by type, to then be collected, pass the inspection to eliminate contaminating elements and inappropriate types of plastic, crushing and washing, followed by separation based on density, drying, casting, drainage through thin screens to remove more contaminating elements; cooling and crushing granules, so that they can finally be resold to plastic companies.

Another recycling method is that which aims at energy recovery through thermal processes. As if it were an incineration, with the difference that the energy generated by burning the plastic is reused. The incineration of 1 kg of recycled plastic generates energy equivalent to burning 1 kg of fuel oil. Countries like Japan and the USA already have hundreds of thermal plants powered by plastic. Legislation also plays a key role in recycling, for example, in January 2018, the European Commission adopted a strategy that proposes to make all plastic packaging reusable or recyclable by 2030 and to reduce the consumption of single-use plastics and MPs. Also in 2018, the European Parliament approved a report that welcomes the European Commission's proposal [67].

Therefore, it is noted the importance of industries to adapt to the development of life cycle assessment (LCA) and related approaches can provide essential guidance to identify the environmentally preferable alternative. Whereas, LCA studies all environmental aspects in order to mitigate, prevent or correct the steps in the preparation of a product, being a guide to manage the preservation of natural resources, optimize recycling and the development of new services and materials, in addition to surveying the potential impacts over the entire life cycle of a product, from the removal of raw materials to disposal [68, 69].

When it comes to the waste generated in greater expansion as doctors, according to the LCA, the incineration of these residues together with the recovery of residual heat is an option that allows the chemical energy content of plastics to be recovered for useful purposes. An LCA with sensitivity analysis of the efficiency of heat recovery confirms that environmental impacts are minimized by maximizing energy recovery. However, there are obstacles to the widespread use of heat recovery incineration. Public concerns about emissions of trace amounts of dioxins and furans can become problematic.

Through Table 1, it can be concluded that the most environmentally friendly material is TNT, given the decomposition time. Derivatives of synthetic rubber, like gloves, take a longer time, but compared to plastics, nothing beats the amount of years for degradability. It is estimated that more than 200 years are needed to degrade a disposable mask, especially if it is discarded in the wrong place or with greater difficulty of access for withdrawal, as in oceans. Emphasizing that, often, the degradation and breakage of this plastic will result in the generation of micro and nano plastics, which make it even more difficult to eliminate the environment.

5.1 Materials, Methodologies and Technologies to Mitigate the Impacts of Waste, Especially in the Pandemic

In view of all the issues already discussed throughout the text, demonstrations of solutions, materials, methodologies and / or technologies already studied, under study or being applied, are necessary to minimize the impacts generated by the COVID pandemic19. Figure 4 shows some of these mitigation actions.

Figure 4 demonstrates some of the main alternatives to mitigate the most seen impacts of the pandemic, especially with regard to the generation of waste.

Some countries already have laws that prohibit the use of single-use plastic bags. Others did not prohibit, but applied taxes on these bags, and unfortunately, there are many countries that allowed free distribution in shops and supermarkets. Faced with the pandemic, in some places, the industry took the opportunity to lift the ban on disposable bags, with claims of viral proliferation, although likewise single-use plastic may still contain viruses and bacteria [70]. The ideal would be the adoption of the use of returnable bags, made of reinforced fabric, suitable for reuse with several purchases, which can also be washed and sanitized after use, thus, there is no excessive generation of plastic bags that may be discarded incorrectly.

Decreasing global subsidies for fossil fuels is emerging, as they contribute to the production of large quantities of greenhouse gas emissions. In addition to GHGs, there is air pollution and, consequently, global warming. The UN also reports that fossil fuel subsidies may also be contributing to COVID-19 mortality rates, based on the link between air pollution, respiratory diseases and the severity of COVID-19 infections [24].

As for the search for solutions for the excessive disposal of PPE, some researchers demonstrate the efficiency of UV rays in the sterilization process, so that they can then reuse some equipment. Bands with a wavelength between 200 and 300 nm can

Table 1 Materials most used in the pandemic, compositions and degradation time

Materials	Composition	Degradation time
Nitrile glove	Acrylonitrile butadiene rubber	Indeterminate
Latex Glove	Natural rubber	Up to 20 years
Vinyl Glove	Vinyl resin	400 years
Disposable plastic glove	High density polyethylene	Up to 500 years
Disposable mask	Non-woven fabric for medical and hospital use (TNT / Nonwoven) Polypropylene, Nontoxic	6 months to 200 years
PFF2 Mask	Outer layer of synthetic polypropylene fiber; middle layer of structural synthetic fibers; synthetic fiber filter layer with electrostatic treatment, inner layer of facial contact synthetic fiber	200 to 600 years
Fabric mask	Cotton or synthetic fabrics	Cotton fabric: 10–20 years Synthetic fabric: 100–300 years
Disposable cap	Non-woven fabric Synthetic, 100% Polypropylene	6 months to 1 year

(continued)

damage the structure of DNA and RNA of several microorganisms, thus inhibiting protein synthesis. There is also the use of chlorine for disinfection, either by jet or washing, but, comparing them, the investment and operating costs of UV disinfection are significantly lower, in addition to not generating aggressive chemical residues. However, there are reports that disinfection with UVC is sometimes unsatisfactory,

Table 1 (continued)

Materials	Composition	Degradation time
Disposable sneaker	Non-woven fabric Synthetic, 100% Polypropylene	6 months to 1 year
Disposable aprons	Non-woven fabric Synthetic, 100% Polypropylene	6 months to 1 year
Face shield	PET; ABS; Polyurethane	200–600 years
Protective plastic for common objects	Polyvinyl chloride	200–600 years

Source Authors, 2021

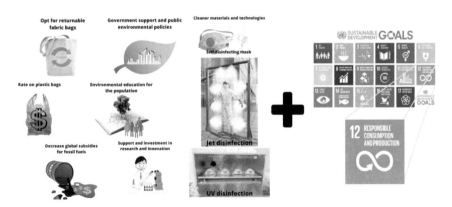

Fig. 4 Actions to mitigate the impacts generated by the COVID pandemic 19. *Source* Authors, 2021

since the depth of penetration is inadequate and there are risks to occupational health [71].

Much of what is known today about COVID-19, such as forms of prevention, care, impacts, development of medications and vaccines, and in addition to the health area, studies that also involve social, economic, environmental impacts, the development of new technologies to mitigate impacts, all these lines of research are developed by countless researchers, scientists from all over the world, who, in the last year, have demonstrated how important and necessary they are, especially in the face of global crises. Therefore, there is a need for greater investment in the scientific area, whether for purchase equipment, tests, locomotion, financial support to institutions and researchers, and technological support and recognition [72].

There are researchers ascribing as possible reuse of the masks as a complementary supply and as a basis for paving streets [66]. Another solution found was the development of respiratory masks that self-disinfect using a heating system using electricity, such as a USB cable. Jet disinfection of chemicals in PPE is also feasible. These and other solutions that have emerged, are seeking to optimize, reuse and avoid the excessive use and disposal of PPE.

The development of new materials, biodegradable, of natural origin or that do not require so many resources, becomes attractive, and can be strong candidates as substitutes for petroleum products, either for making PPE, as for packaging products, food and more goods that depend on plastic and derivatives. Biopolymers are the focus of several studies when it comes mainly to environmental impacts [73].

A biodegradable polymer is composed of high molecular weight that degrades into low molecular weight compounds due to the action of microorganisms or enzymes. Like petroleum-based plastics, biodegradables can be recycled or incinerated, but they are not commonly recycled, as they are seen as contaminants in the recycling system. As they can be degraded by microorganisms, this allows the alternative management of end of life, through industrial and residential composting, facilitating the development of a circular economy. These biopolymers can be widely applied in packaging, food bags, production of objects and more.

Of the highly explored biopolymers globally, polyhydroxyalkanoates (PHA) and polylactic acid (PLA) stand out, PHAs are polymers synthesized by many bacteria as a response to stress during nutrient deprivation and excess carbon. PHA is produced through microbial fermentation processes, where these microorganisms remain as an energy source for survival, therefore, they are present in the cytoplasm of cells in the form of granules and subsequently PHA is extracted by lysis of these cells [74, 75]. Polylactic acid (PLA) is a polyester that has already represented 13.9% of the global bioplastic production capacity. PLA monomers are produced in the process of microbial fermentative and later chemically polymerized. PLA has mechanical and thermal properties similar to those of widely used polymers, such as PET [75, 76].

The applicability of starch and celluloses becomes attractive, after all, starch is a bio-based biodegradable polysaccharide synthesized by most plants through photosynthesis. Starch is an abundantly available, renewable and inexpensive biopolymer. Due to the abundance, low cost and biodegradability of starch, it becomes highly interesting for packaging, being one of the first bioplastics to be commercialized. Starch-based polymers have already represented 21.3% of global bioplastic production capacities. However, given its high hydrophilicity, its processability as a thermoplastic polymer is strongly affected. The most found sources of starch are potatoes, corn, wheat, tapioca, rice, and peas [43, 51].

Cellulose is the most abundant natural biopolymer on Earth, predominantly from trees, being a constituent of plant cell walls, so thousands of tons are produced annually. Cellulose is a linear homopolymer, with microfibrils of diameters of approximately 3–4 nm and, subsequently, microfibrils of 10–25 nm in diameter. In addition to being sustainable, renewable, recyclable, non-toxic and producing a low carbon footprint, cellulose has characteristics that allow it to be applied in several areas, such as the reconstitution of skin and bones, such as membranes for the delivery of drugs and cosmetics, stabilizer of emulsions, food packaging. Due to its mechanical, thermal, optical properties, wettability, among others [77].

Finally, an efficient methodology that must be adopted is the mass awareness of the population about the negative environmental impacts that our actions cause through an accessible and environmental education. Especially, with regard to responsible consumption, and at this pandemic moment, the use of fabric masks should be discussed, demonstrating the need to make the correct disposal, avoiding cross-contamination and the generation of waste in inappropriate places.

6 Who is Responsible for Complying with SDG 12?

All objectives are correlated with each other, complementing and aggregating each other. Therefore, when it comes to SDG 12, it becomes inevitable not to comment on the others. Regarding the responsibility for the fulfillment of these goals and the interested parties for the same are shown in Fig. 5.

A large part of these alternatives is the responsibility of government agencies, the support of public policies, and mainly when putting into practice all the discussions already raised regarding the generation of plastic waste, subsidies on fossil fuels, investment in research, education, innovation, technologies and scientific support. The Sustainable Development Goals present a complete agenda, being a guide for the search for a more environmentally responsible, socially equitable and economically fair society. One hundred and ninety-three delegations have signed the commitment to this agenda, now they just have to comply and recognize that the sustainable and the guarantee that the next generations can enjoy the same things as the current one, depends on the glocal union in favor of the fulfillment of the goals and objectives of sustainable development.

As a civil society, it is necessary to rethink our attitudes towards the use of plastic, the disposal of the residues we produce, the separation of the waste as to the type for disposal of recycling or landfills, promoting sustainable behaviors, breaking old habits and adopting new ones. For this, it is important to invest and stimulate scientific research for the development of new materials, analysis of the impact of waste, methodologies that alleviate environmental degradation, in addition to working with environmental education to increase the public's environmental awareness to encourage them to change their lifestyle, consumption patterns and behavior. Awareness about plastic waste and contamination should not be interrupted

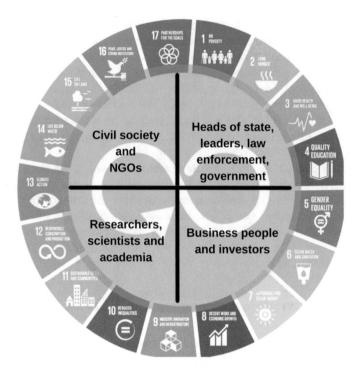

Fig. 5 Stakeholders and allied with the fulfillment of the sustainable development goals. *Source* Authors, 2021

or reversed, as it requires long-term efforts to result in behavioral changes, which can be losses due to interruption or contradictory information [78].

Industries, large investors and business people are essential as regards the parties interested in complying with the SDGs, since they are responsible for the mass production of various materials, consisting of the extraction of raw materials followed by the transformation of it into a product, so that there is distribution to the consumer. Therefore, ecodesign proposals fit into the product design stage, prioritizing materials, technologies and methodologies that do not harm the environment, also associating the analysis of the life cycle of each product, seeking alternatives that mitigate the environmental impact [79]. Therefore, each industry is responsible for the commitment and concern for environmental, social and economic causes, mitigating greenhouse gas emissions, carrying out social projects, campaigns to collect packaging for reuse and recycling, educational announcements on the correct disposal of the product produced, information on the impacts that improper waste abandonment can generate in the short and long term at the local, regional and global levels.

The four axes of civil society, industries, academia and governments are the basis for the fulfillment of the Sustainable Development Goals, therefore competing, the action of all of us, for the construction of a society aligned with sustainable and resilient development.

7   Final Considerations and Future Perspectives

The COVID-19 pandemic has led to historical changes in the norms of our society and in the way people interact. Viral infection, with a significant mortality rate, has impacted all social spheres, social distance has become essential, in addition to the use of masks and hand sanitizers by the population. When it comes to medical care, health professionals redoubled the rigor in the area of individual protection, opting for the use of double gloves, even more sealed masks, excessive exchanges of PPE, and with this, an increase in hospital waste and also an increase in the demand for medical supplies. personal protective equipment by both health agents and the population.

The first pandemic months were uncertain, as it was not known exactly what it was about, what the symptoms and consequences were. In view of this, the best approach was to stop everything that involved people, such as companies, businesses, industries, schools, universities and even some essential services such as garbage collection and separation. Consequently, there was an overload on the system, and a decrease in the correct separation of common and recycled garbage, reducing the percentage of collection of recyclables. In addition, the number of masks and PPE that started to be found in public places, streets, and even exposed in natural environments such as rivers and oceans is significant. All of these issues triggered a warning from the scientific community as to what short and long-term impacts this pandemic can cause on the generation and disposal of waste.

Sustainable Development Goal 12 fits best when studying waste generation. In this chapter, the first goals were highlighted as the main participants in the fight against mismanagement in this critical moment that the world finds itself.

Regarding goal 12.2, it is important to note that good management and the use of natural resources maintain and / or increase the resilience of ecosystems and the benefits they provide. Especially, with regard to the current moment, with the crisis established by COVID-19, they make the search for global goals of sustainability even more dependent on urgent investment decisions to be made by the public and private sectors.

It was also observed that our improper interaction with natural resources can trigger a series of problems. The use of biological resources through our food or through deforestation has several consequences such as the origin of epidemics. As studies have shown, most pathogens move from their wildlife reservoirs to human populations through hunting, the consumption of wild species, the wildlife trade and other forms of contact with wildlife.

Goal 12.3 also showed us the great impact of COVID-19 on the food and agriculture sector and exposed its vulnerabilities. For, as we well know, global food and nutrition security has been threatened since well before the pandemic's arrival. However, in the case of the pandemic, the short-term impacts of COVID-19 required immediate responses to limit the spread of infection through the implementation of health care and containment measures, as the agri-food system is considered a public health instrument. Thus, it must emphasize the importance of good management, of partnerships between countries, mainly, with a focus on developing countries that are historically severely affected by hunger and poverty.

As reported, goals 12.4 and 12.5 become essential guides when it comes to reducing the release of waste into the environment, and the importance of prevention and recycling. These are proven to be the most effective methodologies that guarantee sustainable development and, consequently, compliance with the SDGs. These same goals need support from stakeholders, be they the scientific community, governments, and civil society as a whole. To this end, preserving natural resources, mitigating negative repercussions of the excessive use of fossil derivatives, guaranteeing future generations access to the same resources that current ones have. In addition to ensuring the preservation of natural habitats, in order to prevent the emergence of pandemics with zoonotic origin like the current one.

Therefore, the pandemic shows the need for well-designed and efficient management strategies, after all, we are not exempt from catastrophic occurrences that show the importance of having a balance between human beings and the environment. Thus, responsible consumption and production in favor of sustainable development should continue to be the focus of scientific research.

Acknowledgements This study was conducted by the Centre for Sustainable Development (Greens), from the University of Southern Santa Catarina (Unisul) and Ânima Institute—AI, in the context of the project BRIDGE—Building Resilience in a Dynamic Global Economy: Complexity across scales in the Brazilian Food-Water-Energy Nexus; funded by the Newton Fund, Fundação de Amparo à Pesquisa e Inovação do Estado de Santa Catarina (FAPESC), Coordenação de Aperfeiçoamento de Pessoal de Nível superior (CAPES), National Council for Scientific and Technological Development (CNPq) and the Research Councils United Kingdom (RCUK).

References

1. Platto S, Wang Y, Zhou J, Carafoli E (2020) History of the COVID-19 pandemic: origin, explosion, worldwide spreading. Biochem Biophys Res Commun. https://doi.org/10.1016/j.bbrc.2020.10.087
2. Shereen MA, Khan S, Kazmi A, Bashir N, Siddique R (2020) COVID-19 infection: origin, transmission, and characteristics of human coronaviruses. J Adv Res 24:91–98
3. Córdoba-Aguilar A, Ibarra-Cerdeña CN, Castro-Arellano I, Suzan G (2021) Tackling zoonoses in a crowded world: Lessons to be learned from the COVID-19 pandemic. Acta Trop 214:105780. https://doi.org/10.1016/j.actatropica.2020.105780
4. Parthasarathy P, Vivekanandan S (2020) An extensive study on the COVID-19 pandemic, an emerging global crisis: risks, transmission, impacts and mitigation. J Infect Public Health 14(2):249–259
5. ur Rehman MF, Fariha C, Anwar A, Shahzad N, Ahmad M, Mukhtar S, Haque MFU (2020) Novel coronavirus disease (COVID-19) pandemic: a recent mini review. Comput Struct Biotechnol J 19:612–623
6. Rowan NJ, Laffey JG (2020) Challenges and solutions for addressing critical shortage of supply chain for personal and protective equipment (PPE) arising from Coronavirus disease (COVID19) pandemic: case study from the Republic of Ireland. Sci Total Environ 10:138532. https://doi.org/10.1016/j.scitotenv.2020.138532

7. Hicks A, Temizel-Sekeryan S, Kontar W, Ghamkhar R, Morris MR (2020) Personal respiratory protection and resiliency in a pandemic, the evolving disposable versus reusable debate and its effect on waste generation. Resour Conser Recycl 168:105262

8. Morato MM, Pataro I, Americano da Costa MV, Normey-Rico JE (2020) A parametrized nonlinear predictive control strategy for relaxing COVID-19 social distancing measures in Brazil. ISA transactions, S0019-0578(20)30531-0. Adv online publ. https://doi.org/10.1016/j.isatra.2020.12.012

9. Sen-Crowe B, McKenney M, Elkbuli A (2020) Social distancing during the COVID-19 pandemic: Staying home save lives. Am J Emerg Med 38(7):1519–1520

10. Tonne C (2021) Lessons from the COVID-19 pandemic for accelerating sustainable development. Environ Res 193(November):110482. https://doi.org/10.1016/j.envres.2020.110482

11. Severo EA, Guimarães JCFD, Dellarmelin ML (2021) Impact of the COVID-19 pandemic on environmental awareness, sustainable consumption and social responsibility: evidence from generations in Brazil and Portugal. J Clean Prod 286(2021):124947. https://doi.org/10.1016/j.jclepro.2020.124947

12. Quatrini S (2021) Challenges and opportunities to scale up sustainable finance after the COVID-19 crisis: Lessons and promising innovations from science and practice. Ecosyst Serv 48:101240. https://doi.org/10.1016/j.ecoser.2020.101240

13. Filho WL, Shiel C, Paço A, Mifsud M, Ávila LV, Brandli LL, Molthan-Hil P, Azeiteiro UM, Vargas VR, Caeiro S (2019) Sustainable development goals and sustainability teaching at universities: falling behind or getting ahead of the pack? J Clean Prod 232(2019):285–294. https://doi.org/10.1016/j.jclepro.2019.05.309

14. Kulkarni BN, Anantharama V (2020) Repercussions of COVID-19 pandemic on municipal solid waste management: challenges and opportunities. Sci Total Environ 743:1–37. https://doi.org/10.1016/j.scitotenv.2020.140693

15. Silva ALP, Prata JC, Walker TR, Campos D, Duarte AC, Soares AMVM, Barcelò D, Rocha-Santos T (2020) Rethinking and optimising plastic waste management under COVID-19 pandemic: Policy solutions based on redesign and reduction of single-use plastics and personal protective equipment. Sci Total Environ 742:140565. https://doi.org/10.1016/j.scitotenv.2020.140565

16. Shakil MH, Munim ZH, Tasnia M, Sarowar S (2020) COVID-19 and the environment: a critical review and research agenda. Sci Total Environ 745:141022. https://doi.org/10.1016/j.scitotenv.2020.141022

17. Sharma HB, Vanapalli KR, Cheela VS, Ranjan VP, Jaglan AK, Dubey B, Goel S, Bhattacharya J (2020) Challenges, opportunities, and innovations for effective solid waste management during and post COVID-19 pandemic. Resour Conserv Recycl 162(May):105052. https://doi.org/10.1016/j.resconrec.2020.105052

18. Bayulken B, Huisingh D, Fisher PMJ (2021) How are nature based solutions helping in the greening of cities in the context of crises such as climate change and pandemics? A comprehensive review. J Clean Prod 288:125569. https://doi.org/10.1016/j.jclepro.2020.125569

19. Managing ASFC, Ashton W, Moreau V, Tseng ML (2018) Sustainable management of natural resources toward sustainable development goals. Resour Conserv Recycl 136(May):335–336. https://doi.org/10.1016/j.resconrec.2018.04.019

20. Tahir MB, Batool A (2020) COVID-19: Healthy environmental impact for public safety and menaces oil market. Sci Total Environ 740:140054. https://doi.org/10.1016/j.scitotenv.2020.140054

21. Juanwen Y, Quanxin W, Jinlong L (2012) Understanding indigenous knowledge in sustainable management of natural resources in China. Taking two villages from Guizhou Province as a case. Forest Policy Econ 22(9):47–52. https://doi.org/10.1016/j.forpol.2012.02.012

22. Sangha KK, Maynard S, Pearsond J, Dobriyale P, Badolae R, Hussain SA (2019) Recognising the role of local and indigenous communities in managing natural resources for the greater public benefit: Case studies from Asia and Oceania region. Ecosyst Serv 39(June):100991. https://doi.org/10.1016/j.ecoser.2019.100991

23. United Nations Development Programme (2015) Sustainable development goals. United Nations Dev Programme. Available at: www.undp.org. Accessed on: 03 May 2021.
24. United Nations (2015) Ensure sustainable consumption and production patterns. United Nations. Available at: https://sdgs.un.org/goals/goal12. Accessed on: 03 May 2021.
25. Halonen JI, Erhola M, Furman E, Haahtela T, Jousilahti P, Barouki R, Bergman Å, Billo NE, Fuller R, Haines A, Kogevinas M (2021) A call for urgent action to safeguard our planet and our health in line with the Helsinki declaration. Environ Res 193:1–8. https://doi.org/10.1016/j.envres.2020.110600
26. Brilha J, Gray M, Pereira DI, Pereira P (2018) Geodiversity: An integrative review as a contribution to the sustainable management of the whole of nature. Environ Sci Policy 86(January):19–28. https://doi.org/10.1016/j.envsci.2018.05.001
27. Everard M, Johnstonb P, Santillo D, Staddon C (2020) The role of ecosystems in mitigation and management of Covid-19 and other zoonoses. Environ Sci Policy 111(May):7–17. https://doi.org/10.1016/j.envsci.2020.05.017
28. You M (2020) Changes of China's regulatory regime on commercial artificial breeding of terrestrial wildlife in time of COVID-19 outbreak and impacts on the future. Biol Conserv J 250(March):108756. https://doi.org/10.1016/j.biocon.2020.108756
29. Lahrich S, Laghrib F, Farahi A, Bakasse M, Saqrane S, El Mhammedi MA (2021) Review on the contamination of wastewater by COVID-19 virus: impact and treatment. Sci Total Environ 751:142325. https://doi.org/10.1016/j.scitotenv.2020.142325
30. Fernandez D, Gine-Vazquez I, Liu I, Yucel R, Ruscone MN, Morena M, García VG, Haro JM, Pan W, Tyrovolas S (2021) Are environmental pollution and biodiversity levels associated to the spread and mortality of COVID-19? A four-month global analysis. Environ Pollut 271:116326. https://doi.org/10.1016/j.envpol.2020.116326
31. CBD (2020) Global biodiversity outlook 5 summary. Secretariat of the convention on biol diversity (2020) global biodiversity outlook 5. Montreal. Available on: https://www.cbd.int/gbo/. Accessed on: 03 may 2021
32. Sarkodie SA (2021) Environmental performance, biocapacity, carbon & ecological footprint of nations: drivers, trends and mitigation options. Sci Total Environ 751:141912. https://doi.org/10.1016/j.scitotenv.2020.141912
33. GFN (2020) Open data plataform. Global footprint network. Available at: https://data.footprintnetwork.org/#/. Accessed on: 03 May 2021
34. Hou D, Bolan NS, Tsang DCW, Kirkham MB, O'Connor D (2020) Sustainable soil use and management: An interdisciplinary and systematic approach. Sci Total Environ J 729:138961. https://doi.org/10.1016/j.scitotenv.2020.138961
35. Zhu Y, Ali SH, Xu D, Cheng J (2021) Mineral supply challenges during the COVID-19 pandemic suggest need for international supply security mechanism. Resour Conserv Recycl 165(October):105231. https://doi.org/10.1016/j.resconrec.2020.105231
36. Hilson G, Bockstael SV, Sauerwein T, Hilson A, McQuilken J (2021) Artisanal and small-scale mining, and COVID-19 in sub-Saharan Africa: A preliminary analysis. World Dev 139:105315. https://doi.org/10.1016/j.worlddev.2020.105315
37. Huo C, Dar AA, Nawaz A, Hameed J, Albashar G, Pan B, Wang C (2021) Groundwater contamination with the threat of COVID-19: Insights into CSR theory of Carroll's pyramid. J King Saud Univ Sci 33(2):101295. https://doi.org/10.1016/j.jksus.2020.101295
38. Antwi SH, Getty D, Linnane S, Rolston A (2021) COVID-19 water sector responses in Europe: A scoping review of preliminary governmental interventions. Sci Total Environ 762:143068. https://doi.org/10.1016/j.scitotenv.2020.143068
39. Kalbuscha A, Henninga E, Brikalskia MP, Luca FV, AC, Konrath (2020) Impact of coronavirus (COVID-19) spread-prevention actions on urban water consumption. Resour Conserv Recycl 163(May):105098. https://doi.org/10.1016/j.resconrec.2020.105098
40. Chofreh AG, Goni FA, Kleme JJ, Moosavi SMS, Davoudi M, Zeinalnezhad M (2021) Covid-19 shock: development of strategic management framework for global energy. Renew Sustain Energy Rev 139(June):110643. https://doi.org/10.1016/j.rser.2020.110643

41. Jiang P, Fan YV, Kleme JJ (2021) Impacts of COVID-19 on energy demand and consumption: Challenges, lessons and emerging opportunities. Appl Energy 285(December):116441. https://doi.org/10.1016/j.apenergy.2021.116441

42. Elavarasan RM, Shafiullah G, Raju K, Mudgal V, Arife MT, Jamal T, Subramanian S, Balaguru VSS, Reddy KS, Subramaniam U (2020) COVID-19: impact analysis and recommendations for power sector operation. Appl Energy 279:115739. https://doi.org/10.1016/j.apenergy.2020.115739

43. Dutta A, Das D, Jana RK, Vo XV (2020) COVID-19 and oil market crash: Revisiting the safe haven property of gold and Bitcoin. Resour Policy 69(April):101816. https://doi.org/10.1016/j.resourpol.2020.101816

44. Zanocco C, Flora J, Rajagopal R, Boudet H (2021) Exploring the effects of California's COVID-19 shelter-in-place order on household energy practices and intention to adopt smart home technologies. Renew Sustain Energy Rev 139(December):110578. https://doi.org/10.1016/j.rser.2020.110578

45. Nchanji EB, Lutomia CK (2021) Regional impact of COVID-19 on the production and food security of common bean smallholder farmers in Sub-Saharan Africa: implication for SDG's. Glob Food Sec 29(January):100524. https://doi.org/10.1016/j.gfs.2021.100524

46. Han S, Roy PK, Hossain MI, Byun K-H, Choi C, Ha S-D (2021) COVID-19 pandemic crisis and food safety: implications and inactivation strategies. Trends Food Sci Technol 109(January):25–36. https://doi.org/10.1016/j.tifs.2021.01.004

47. Stephens EC, Martin G, van Wijk M, Timsina J, Snow V (2020) Editorial: Impacts of COVID-19 on agricultural and food systems worldwide and on progress to the sustainable development goals. Agric Syst 183:102873. https://doi.org/10.1016/j.agsy.2020.102873

48. Fan S, Teng P, Chew P, Smith G, Les C (2021) Food system resilience and COVID-19: lessons from the Asian experience. Glob Food Sec 28:100501. https://doi.org/10.1016/j.gfs.2021.100501

49. Iese V, Wairiu M, Hickey GM, Ugalde D, Salili DH, Walenenea J Jr, Tabe T, Keremama M, Teva C, Navunicagi O, Fesaitu J (2021) Impacts of COVID-19 on agriculture and food systems in Pacific Island countries (PICs): Evidence from communities in Fiji and Solomon Islands. Agric Syst 190:103099. https://doi.org/10.1016/j.agsy.2021.103099

50. Sunny AR, Sazzad SA, Prodhan SH, Ashrafuzzaman M, Datta GC, Sarker AK, Rahman M, Mithun MH (2021) Assessing impacts of COVID-19 on aquatic food system and small-scale fisheries in Bangladesh. Mar Policy 126:104422. https://doi.org/10.1016/j.marpol.2021.104422

51. Panzone LA, Larcom S, She PW (2021) Estimating the impact of the first COVID-19 lockdown on UK food retailers and the restaurant sector. Glob Food Sec 28(October):100495. https://doi.org/10.1016/j.gfs.2021.100495

52. Kim J, Kim J, Wang Y (2021) Uncertainty risks and strategic reaction of restaurant firms amid COVID-19: evidence from China. Int J Hosp Manag 92(October):102752. https://doi.org/10.1016/j.ijhm.2020.102752

53. Brewer P, Sebby AG (2021) The effect of online restaurant menus on consumers' purchase intentions during the COVID-19 pandemic. Int J Hosp Manag 94(October):102777. https://doi.org/10.1016/j.ijhm.2020.102777

54. Byrd K, Her E, Fan A, Almanza B, Liu Y, School SL (2021) Restaurants and COVID-19: What are consumers' risk perceptions about restaurant food and its packaging during the pandemic? Int J Hosp Manag 94(June):102821. https://doi.org/10.1016/j.ijhm.2020.102821

55. Principato L, Secondi L, Cicatiello C, Mattia G (2021) Caring more about food: The unexpected positive effect of the Covid-19 lockdown on household food management and waste. Socio-Econ Plan Sci 2020:100953. https://doi.org/10.1016/j.seps.2020.100953

56. Aldaco R, Hoehn D, Laso J, Margallo M, Ruiz-Salmón J, Cristobal J, Kahhat R, Villanueva-Rey P, Bala A, Batlle-Bayer L, Fullana-i-Palmer P, Irabien A, Vazquez-Rowe I (2020) Food waste management during the COVID-19 outbreak: a holistic climate, economic and nutritional approach. Sci Total Environ 742:140524. https://doi.org/10.1016/j.scitotenv.2020.140524

57. Rume T, Islam SMD-U (2020) Environmental effects of COVID-19 pandemic and potential strategies of sustainability. Heliyon 6(9):e04965. https://doi.org/10.1016/j.heliyon.2020.e04965
58. Oli AN, Ekejindu CC, Adje DU, Ezeobi I, Ejiofor OS, Ibeh CC, Ubajaka CF (2016) Healthcare waste management in selected government and private hospitals in Southeast Nigeria. Asian Pac J Trop Biomed 6(1):84–89. https://doi.org/10.1016/j.apjtb.2015.09.019
59. Ammendolia J, Saturno J, Brooks AL, Jacobs S, Jambeck JR (2021) An emerging source of plastic pollution: environmental presence of plastic personal protective equipment (PPE) debris related to COVID-19 in a metropolitan city. Environmental Pollution 269:116160
60. HEALTH (2020) Secretariat of learn how to properly dispose and sanitize such protective masks. Available at: https://saude.rs.gov.br/saiba-como-descartar-e-higienizar-corretamente-as-mascaras-de-protecao. Accessed on: 26 Jan. 2021
61. PINHEIRO, Layse T (2018) Carbon dioxide and methane fluxes from an amazon dump. Advisor: José Henrique Cattanio. 2018. 111 f. Dissertation (Master's Degree in Environmental Sciences) - Institute of Geosciences, Federal University of Pará, Museu Paraense Emílio Goeldi, Brazilian Agric Res Corporation, Belém. Available at: http://repositorio.ufpa.br/jspui/handle/2011/11045
62. Golwala H, Zhang X, Iskander SM, Smith AL (2021) Solid waste: an overlooked source of microplastics to the environment. Sci Total Environ 769:144581
63. Cheng Y, Song W, Tian H, Zhang K, Li B, Du Z, Zhang W, Wang J, Wang J, Zhu L (2021) The effects of high-density polyethylene and polypropylene microplastics on the soil and earthworm Metaphire guillelmi gut microbiota. Chemosphere 267:129219. https://doi.org/10.1016/j.chemosphere.2020.129219
64. Dharmaraj S, Ashokkumar V, Hariharan S, Manibharathi A, Show PL, Chong CT, Ngamcharussrivichai C (2021) The COVID-19 pandemic face mask waste: a blooming threat to the marine environment. Chemosphere 272:129601
65. Peng X, Cao J, Xie B, Duan M, Zhao J (2020) Evaluation of degradation behavior over tetracycline hydrochloride by microbial electrochemical technology: Performance, kinetics, and microbial communities. Ecotoxicol Environ Saf 188:109869
66. Saberian M, Li J, Kilmartin-Lynch S, Boroujeni M (2021) Repurposing of COVID-19 single-use face masks for pavements base/subbase. Sci Total Environ 769:145527
67. Guerranti C, Martellini T, Perra G, Scopetani C, Cincinelli A (2019) Microplastics in cosmetics: Environmental issues and needs for global bans. Environ Toxicol Pharmacol 68:75–79
68. Silva RR, de Lima Rodrigues FTR (2015) Análise do ciclo de vida e da logística reversa como ferramentas de gestão sustentável: o caso das embalagens PET. Iberoamerican J Ind Eng 7(13):44–58
69. Vista HAB, dos Santos MR, Shibao FY (2015) Produto Sustentável: Equipamento de Proteção Individual Fabricado com Plástico Verde. Revista de Gestão Ambiental e Sustentabilidade: GeAS 4(1):59–71
70. Zambrano-Monserrate MA, Ruano MA, Sanchez-Alcalde L (2020) Indirect effects of COVID-19 on the environment. Sci Total Environ 728:138813
71. Wang J, Li W, Yang B, Cheng X, Tian Z, Guo H (2020) Impact of hydrological factors on the dynamic of COVID-19 epidemic: a multi-region study in China. Environ Res 198:110474. https://doi.org/10.1016/j.envres.2020.110474
72. Serafim MP, Dias RDB (2020) The importance of science and public universities in solving social problems. Evaluation: J Eval Higher Educ (Campinas), 25(1):1–4
73. Yin GZ, Yang XM (2020) Biodegradable polymers: a cure for the planet, but a long way to go. J Polym Res 27(2):1–14
74. Fortuna ALL (2020) Environmental impacts of plastics: biopolymers as an alternative to reduce the accumulation of flexible polypropylene packaging in the environment. Available at: https://pantheon.ufrj.br/bitstream/11422/12581/1/ALLFortuna.pdf. Accessed on: 03 May 2021
75. RameshKumar S, Shaiju P, O'Connor KE (2020) Bio-based and biodegradable polymers-State-of-the-art, challenges and emerging trends. Curr Opin Green Sustain Chem 21:75–81

76. Oliveira ACS, Borges SV (2020) Poly (lactic acid) applied to food packaging: a review. Electron J Mater Processes, 15(1)
77. Gan I, Chow WS (2018) Antimicrobial poly (lactic acid)/cellulose bionanocomposite for food packaging application: a review. Food Packag Shelf Life 17:150–161
78. Silva ALP, Prata JC, Walker TR, Duarte AC, Ouyang W, Barcelò D, Rocha-Santos T (2021) Increased plastic pollution due to COVID-19 pandemic: Challenges and recommendations. Chem Eng J 405:126683. https://doi.org/10.1016/j.cej.2020.126683
79. Borchardt M, Wendt MH, Pereira GM, Sellitto MA (2011) Redesign of a component based on ecodesign practices: environmental impact and cost reduction achievements. J Clean Prod 19(1):49–57
80. Azhar S, Shahzad A, Lawler J, Mahmoud KA, Lee DS, Ali N, Bilal M, Rasool K (2021) Unprecedented environmental and energy impacts and challenges of COVID-19 pandemic. Environ Res 193(November):110443. https://doi.org/10.1016/j.envres.2020.110443
81. Wang J, Shen J, Ye D, Yan X, Zhang Y, Yang W, Li X, Wang J, Zhang L, Pan L (2020) Disinfection technology of hospital wastes and wastewater: Suggestions for disinfection strategy during coronavirus Disease 2019 (COVID-19) pandemic in China. Environ Pollut 262:114665

Traffic Incidents During the COVID-19 Pandemic: A Step Towards Meeting the Sustainable Development Goals

Marina Leite de Barros Baltar, Victor Hugo Souza de Abreu, Glaydston Mattos Ribeiro, and Fábio Ramos

Abstract Currently, efforts have been made to develop studies that seek to ascertain the impact of the blockage implemented to combat the spread of the coronavirus (COVID-19) in various economy sectors such as services, commerce, industry, civil construction, transportation and agriculture. Regarding the transport sector specifically, the blocks implemented in several countries around the world, to contain the virus, resulted in a significant reduction in the number of trips between points of origin and destination, with only the essential ones advised. This reduction should also impact on congestion and the number of traffic incidents. Thus, this chapter seeks to test the hypothesis that isolation, due to COVID-19, resulted in a significant drop in the occurrence of traffic incidents, through the performance of statistical analyzes. In addition, this possible reduction is discussed as being an important strategy to achieve several Sustainable Development Goals (SDGs). We used traffic incidents data in the Rio de Janeiro city made available by Traffic Engineering Company of the city. The results obtained through various statistical methods indicate that although the accident profile has not undergone significant changes, there was a sharp drop in the number of calls for traffic incidents, which can positively impact nine SDGs and 23 targets of the 2030 Agenda. It is also worth mentioning that although the blocking state is not permanent, the findings found can be used to improve the potential for mitigating traffic incidents in the future and help in a future planning of incident management.

M. L. de Barros Baltar (✉) · V. H. S. de Abreu · G. M. Ribeiro
Transport Engineering Program (PET), Alberto Luiz Coimbra Institute for Graduate Studies and Research in Engineering (COPPE), Federal University of Rio de Janeiro (UFRJ), Rio de Janeiro, Brazil
e-mail: mabaltar@pet.coppe.ufrj.br

V. H. S. de Abreu
e-mail: victor@pet.coppe.ufrj.br

G. M. Ribeiro
e-mail: glaydston@pet.coppe.ufrj.br

F. Ramos
Mathematical Institute, Federal University of Rio de Janeiro (UFRJ), Rio de Janeiro, Brazil
e-mail: framos@ufrj.br

© The Author(s), under exclusive license to Springer Nature Singapore Pte Ltd. 2021
S. S. Muthu (ed.), *COVID-19*, Environmental Footprints and Eco-design
of Products and Processes, https://doi.org/10.1007/978-981-16-3860-2_3

73

Keywords Incident management · Transportation · Traffic incidents · COVID-19 · Pandemic · Statistical methods · Sustainable development · Sustainable development goals

1 Introduction

To reduce the effects of COVID-19 pandemic, a fast-spreading disease through respiratory droplets, several countries sought to implement travel restrictions for population [1], with the temporary suspension of most face-to-face activities in favor of virtual and remote activities, to reduce the chances of contracting and spreading the virus [2]. Thus, at the end of March 2020, half of the world population was under some form of blockage [3], closing part of the industrial activities globally [4].

Among many other sectors, transportation is the sector most affected by the pandemic [4], because there was a significant reduction in the number of trips between points of origin and destination due to the lockdowns [1, 5], including commuting and recreational travels [6], with essential travel only being recommended by health and public safety professionals.

COVID-19 pandemic has provided a severe negative impact on human health, on global economy and, consequently, on freight and passenger transportation [5]. However, it also reduces the pollution due to limited social and economic activities [4, 7, 8]. Studies present evidence for a link between the global decline in transportation and the reduction of environmental exposure to various air pollutants [4, 7, 9].

Besides, it is believed that blockages in the transport sector also led to a congestion reduction, due to the lower number of vehicles circulating on the roads [1], and to a reduction in the number of traffic incidents, which can be defined as any non-periodic event that decreases the road capacity or generates an abnormal increase in demand [10]. These incidents cause a intensify traffic congestion, leading to delays in travel times, excessive fossil fuel consumption, increased environmental, visual and noise pollution, among others [11–14].

In relation to accidents and running over, for example, it is observed that, generally, as economic activity decreases, travel decreases and drivers are exposed to a lower collision risk [15]. In fact, during the COVID-19 pandemic, fewer vehicles were observed on the roads [1, 16] and fewer accidents were identified [2, 15, 17, 18]. In addition, the existence of an incident can lead to the occurrence of secondary incidents, aggravating the impacts.

Efforts are currently being made to identify the trends that are presented in the incidence of motor vehicle collisions and vehicle-related injuries, specifically during the COVID-19 pandemic. However, it is highlighted that there is a lack of studies related to the analysis of incidents in general. The reason is that, during the literature review, we identified that the papers that address the theme deal specifically with the analysis of traffic accidents.

Thus, this chapter seeks to test the hypothesis that the isolation due to COVID-19 significantly resulted in declines in the occurrence of traffic incidents in general (including in addition to accidents and running over, the broken-down vehicles that represent a significant portion of these incidents), through a statistical analysis, taking the Rio de Janeiro city, Brazil, as a case study. The reduction in traffic incidents leads to a reduction of the negative impacts, such as increasing of the CO_2 emissions, increasing of the traffic congestion and a reduction of the population life quality.

It is worth mentioning that this chapter is aligned with the 2030 Agenda for Sustainable Development and its 17 Sustainable Development Goals (SDGs) [19], among which the following stand out: SDG 3 (Good Health and Wellbeing), SDG 9 (Industry, Innovation and Infrastructure), SDG 11 (Sustainable Cities and Communities) and SDG 13 (Climate Action). For example, one of the targets of SDG 3 (specifically the Target 3.6) aimed to decrease the global number of road deaths and injuries by 2020. The simple inclusion of such an ambitious target, related to road safety, represents the need to better traffic management with the aim of reducing traffic incidents.

To compare the scenarios before and during the lockdown, carried out to combat the spread of COVID-19, data related to traffic incidents are used which were provided by the Traffic Engineering Company of the Rio de Janeiro city (Companhia de Engenharia de Tráfego da Cidade do Rio de Janeiro—CET-Rio, in Portuguese). In addition, statistical tools are used to observe the real reduction in the number of calls and their configuration in different vehicle categories.

Although this chapter is focused on the scenarios before and during the COVID-19 pandemic, as a result of social distance, the demand for travel may drop permanently due to a greater amount of work done in home office, greater purchases online and people may still fear social contact when the rules of social distance are no longer in effect, reducing the number of activities and travels [1].

The remainder of this chapter is organized as follows. Section 2 presents a bibliographic review and Sect. 3 describes the methodology used. Section 4 shows the results and a discussion is presented. Section 5 seeks to discuss how the reducing of traffic incidents can contribute to reaching targets of SDGs. Finally, the final observations and future directions of the study are expressed in Sect. 6.

2 Bibliographic Review

Epidemic and pandemic outbreaks, such as dengue, swine flu, bird flu and, recently, COVID-19, have a severe and versatile impact on society as well as on the economy [5]. COVID-19 started in Wuhan, in the People's Republic of China, and, in January 2020, it spread quickly not only in the surrounding areas, but also throughout China and the outbreak turned an epidemic [8]. At the end of January of 2020, a worldwide public health emergency was declared [28].

In February of 2020, outbreaks began in Iran, Italy and other countries around the globe. Subsequently, the epidemic turned into a pandemic and, at the end of March of 2020, half of the world population was under some form of blockage [3]. COVID-19 spread rapidly, partly due to the increase in globalization and the accessibility of the first epicenter (Wuhan) [20].

In order to slow the spread of COVID-19, governments around the world have decided to impose several temporary restrictions [5]. Thus, increasingly stringent measures were put in place by world governments in an effort, initially, to isolate cases and interrupt the transmission of the virus and, subsequently, to decrease its rate of spread [21], such as close schools, shops, restaurants and bars, ban public events and encourage the work at home [1].

In Brazil, on March 16 of 2020, the state governor declared a public health emergency in the Rio de Janeiro city and partial blockade measures took effect a week later [7]. During the first week of quarantine in Rio de Janeiro, public transportation was reduced by approximately 50% and private vehicles were significantly reduced [22].

The blocking response to COVID-19 caused an unprecedented reduction in the global economy and in the transport activity [9]. Countries around the world are taking steps to reduce the effects, or at least delay them, to deal better with public health and manage better their limited resources [23]. Some countries, such as China, Italy and Spain, have imposed social distance by imposing blockades (in certain regions or in the country as a whole), while other countries, such as Holland, Sweden, the United Kingdom and the USA, have adopted a less strict social distance for through control measures [1].

As a result, government measures (such as travel restrictions and shutdowns of entire sectors of the economy) have been implemented, as well as individual choices to refrain from traveling in order to reduce exposure to other people and the risk of contamination [24]. The confinement of the population is leading to drastic changes in energy use, with expected impacts on pollutant emissions [21]. The reduction in pollutant emissions has a positive impact on air quality and is discussed especially in the context of COVID-19 and its versatile government restrictions during blocking periods [4, 7–9, 25].

As millions of people started working at home or lost the jobs and the government officials implemented unprecedented measures to restrict travel, there was a huge decrease in commuting, along with even more dramatic decreases in recreational travels [6], resulting in a reduction in the traffic volume [5], and, consequently, a reduction of congestion [1] and traffic incidents such as accidents, run overs, mechanical malfunctions, etc.

Following this line, reference [2], for example, sought to identify and explain what trends have been presented in the incidence of motor vehicle collisions and vehicle-related injuries since the beginning of the COVID-19 pandemic. The authors identified that this incidence decreased significantly during the COVID-19 in some states like Florida, New York and Massachusetts and that the creation of improved public transport modes and the use of virtual/remote activities can serve as solutions to reduce vehicle-related collisions and injuries in a long term.

Reference [18] presented California's preliminary data on total and victim collisions (injuries/fatality) on state highways during the COVID-19 pandemic. The authors indicated that there was a reduction of approximately 50% in the total of accidents, between March 1, 2020, and April 30, 2020, compared to the period before the blockade and a similar period in 2019.

Reference [17] similarly presented North Carolina's total collision and victim data for the period of blocking economic activities in combating the spread of COVID-19, from March 15, 2020, to May 16, 2020. Compared to the pre-lock baseline, the total crashes decreased by 50%, fatal crashes decreased by 10%, while crashes curiously increased by 6%, which was attributed to the higher proportion of crashes of a single vehicle during this period.

Thus, it is noted that currently efforts have been made to identify the trends that are presented in the incidence of motor vehicle collisions and vehicle-related injuries, specifically during the COVID-19 pandemic. However, it is highlighted that there is a lack of studies related to the analysis of incidents in general. The reason is that, as can be seen in the studies discussed here, the papers that address the theme deal specifically with the analysis of traffic accidents. Thus, this chapter seeks to address this gap in the literature by considering, in addition to accidents, mechanical faults and flat tires, which represent a significant portion of traffic incidents. In addition, the existence of these events leads to secondary accidents. So, this study seeks to bring the results closer to the possible benefits for achieving the SDGs and their targets of 2030 Agenda.

3 Methodology

As presented in Sect. 1, this chapter aims to analyze a possible change in the profile of attendances to traffic incidents, due to the social isolation imposed by the COVID-19 pandemic, through data collected in an urban area. For comparison of scenarios before and during the pandemic, statistical tools are used in order to observe the real reduction in the number of calls and the different vehicle considered. The methodology developed has four stages: (1) data collection; (2) data treatment (Elbow and K-Means Methods); (3) statistical analysis (parametric and non-parametric hypothesis tests); and (4) correlation with the SDGs and their targets, as can be seen in Fig. 1.

In the first stage, the aim is to collect data referring to calls for traffic incidents, before and during isolation, with the objective of knowing them through information such as time of occurrence of the incident, incident type, vehicle involved and whether it occurred on business day or not. Information about decrees and laws related to social isolation was also analyzed.

In the second stage, the data are processed, initially excluding incomplete or discrepant information. Subsequently, they were divided into working days, weekends and holidays. It should be noted that, due to the functioning of the cities, the traffic of vehicles on working days is different to the one found during weekends. Traffic flow is usually higher due to study and work trips during working days in urban regions. Therefore, it was decided to focus the analysis on working days only. For this, we used K-Means and Elbow Methods.

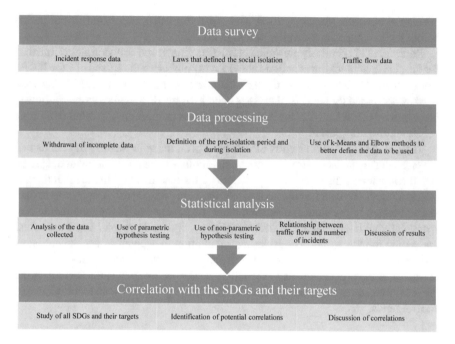

Fig. 1 Methodology adopted

K-Means is a localized optimization method that is sensitive to the selection of the starting position of the clusters' center. This method divides in k clusters the total of observations, but it does not determine the ideal number of clusters (k), so that k needs to be pre-defined. For this, the Elbow Method is used [26]. This method analyzes the percentage of variance explained as a function of the number of clusters and aims to choose a number of clusters so that the addition of one more does not generate better results [27]. The ideal number of clusters is defined by the distortion observed. So, to confirm this difference, the K-means and Elbow methods are used. Thus, it is possible to check whether there is a difference between working days and the other days for incident response data.

After an initial analysis, the data observation periods are defined: one before the pandemic and other after the social isolation. These periods are named Pre-isolation Period (PP) and Insulation Period (IP), respectively. Subsequently, in the third stage, parametric and non-parametric hypothesis tests are performed to analyze the inference between the two periods under study. The first test to be performed is the Z test, which analyzes the difference between the means of the two samples considering the two populations with normal and independent distribution. This test considers the difference between the means, the mean value of the differences for the paired sample data (Δ_0), which is normally equal to 0, the standard deviation and the number of observations, according to Eq. (1).

$$z_0 = \frac{x_1 - x_2 - \Delta_0}{\sqrt{\sigma_1^2/n_1 + \sigma_2^2/n_2}} \tag{1}$$

where:

- x_1 is mean of sample of population 1 (PP);
- x_2 is mean of sample of population 2 (IP);
- Δ_0 is the difference in populations means;
- σ_1 is the standard deviation of population 1 (PP);
- σ_2 is the standard deviation of population 2 (IP);
- n_1 is the size of population 1 (PP); and
- n_2 is the size of population 2 (IP).

It is noteworthy that equivalent values are used in the analysis, due to the difference in the sample sizes of the two periods. These values are calculated using Eq. (2).

$$n = \frac{\sigma_1^2 + \sigma_2^2}{\sigma_1^2/n_1 + \sigma_2^2/n_2} \tag{2}$$

where:

- σ_1 is the standard deviation of population 1 (PP);
- σ_2 is the standard deviation of population 2 (IP);
- n_1 is the size of population 1 (PP); and
- n_2 is the size of population 2 (IP).

The second hypothesis test performed is the nonparametric test, based on the difference between the medians. This test is advantageous in asymmetric distributions since the median is closer to the data mass than the average. For this analysis to become possible, the bootstrap technique is used because of its capability of correcting eventual random errors inherent to the data. From this technique, a sample of the studied population is collected and an estimate of the parameter of interest is calculated, the median, in this case, from the sample data. Then a brief correlation is presented between the traffic flow and the number of incidents attended per day.

In the fourth stage, considering the SDGs and targets of 2030 Agenda, finally we sought to correlate and discuss how each of them can be improved through potential reductions in incidents, supported by the literature review.

4 Case Study

The methodology, described in Sect. 3, is applied to analyze a possible change in the profile of handling traffic incidents, before and during the COVID-19 pandemic, in the Rio de Janeiro city, through data provided by CET-Rio, as previously exposed.

This case study is relevant due to the fact that Brazil is one of the countries with the highest rate of contaminated and dead in the world [28], with a significant drop in number of trips.

Rio de Janeiro is the second state with the highest record of deaths by COVID-19 in Brazil, São Paulo is the first one, according to the Ministry of Health [29]. The capital of these two states are the largest metropolises in the country [30] and the places with the highest concentration of people, which enables a greater spread of the virus.

Thus, in Rio de Janeiro, the main city in the state, several measures were adopted due to COVID-19 to ensure the social distance such as suspension of reversible lanes, closing of trade and change of schedule in services that needed to be kept open, according to Decree Rio n° 47.282 of 21 March 2020 (Rio de Janeiro City Hall, 2020).

This decree caused a major change in the city's routine, significantly reduced the number of trips, which also generates a reduction in the congestion and in the handling of incidents. It is noteworthy that the Rio de Janeiro appears as the second most congested city in Brazil, according to the classification of the Tomtom Report of 2019 [31]. These congestions happen daily and the presence of several non-periodic incidents, cause a reduction in road capacity or an abnormal increase in demand.

Thus, to verify how the blocking measures implemented in the Rio de Janeiro city had an effect in the handling of traffic incidents, 155,476 records were analyzed which occurred between 2015 and May 2020, divided between incident type (broken-down vehicles and accidents, the latter includes running over), characterized by vehicle type involved and time of occurrence. It is worth mentioning that incidents with buses are not analyzed, since the attendance form for this vehicle type changed at the beginning of the isolation period.

To analyze only working days, the Elbow and K-means Methods are used in an integrated manner. Through the first method it is possible to verify which number of clusters (k) brings good results. Thus, as can be seen in Fig. 2, from $k = 3$ it is possible to observe a marked reduction in distortion, concluding that from 3 clusters, the modeling already brings good results. The distortion is the sum of the square distance from each point of the population to its center, in other words, less distortion corresponds a better clustering result. Therefore, this is the number of clusters defined by the K-means Method.

The percentage of Saturdays, Sundays, holidays and working days in each cluster is defined by clustering, carried out by the K-means Method, as shown in Table 1. Working days were divided into two clusters, in which 37.8% were located in Cluster 0 and 48.9% in Cluster 1. In Cluster 1, 54.61% of Saturdays were located. Already 38% of Saturdays were on the same clusters as most Sundays (86.3%) and holidays (90.1%), Cluster 2.

When plotting the profile of the clusters, that is, the average number of calls for incidents per hour, it is observed that Cluster 0, where most Sundays and holidays and part of Saturdays were located, there was a marked reduction in the total number of incidents attended, as shown in Fig. 3a. Cluster 1 and Cluster 2 have similar curves (Fig. 3b and c). However, the number of incidents per hour in Cluster 1, where most

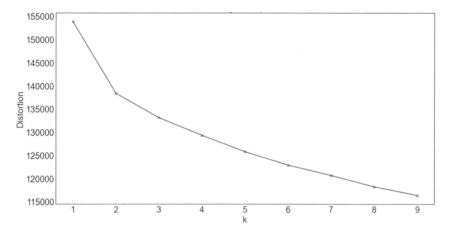

Fig. 2 Result of the Elbow Method for the ideal number of clusters

Table 1 Result of clustering using the K-means method

Period	Clusters		
	0 (%)	1 (%)	2 (%)
Working day	37.80	48.96	13.24
Saturday	7.38	54.61	38.01
Sunday	0.37	13.28	86.35
Holiday	0.00	9.84	90.16

Saturdays were located and the time with the highest number of attendances records 6 incidents in an hour, is less than in Cluster 2, where the hour with the greatest number of attendances recorded 8 incidents. In addition, the peak in the afternoon of Cluster 1 is less than the one in Cluster 2. Thus, Saturday mornings are similar to weekdays because commerce and schools are opened.

Thus, due to this difference in the behavior and the observation of greater linearity in the data on working days, only those days are used for the analysis by vehicle type of attendance to incidents in the pre-isolation period and during the isolation.

Figure 4 shows the number of incidents attended per day throughout the Rio de Janeiro city, from 2015 to May 2020. It shows that, from the moment of isolation, represented by the dashed red line, there was a sharp reduction in the number of daily incident response performed. It is also noted that the reduction in the average number of incident response during the isolation period was 41.6%.

It is also worth mentioning that, despite the significant reduction in the number of incidents, it can be seen in Fig. 5 that the behavior of the occurrence of incidents throughout the day remained similar when comparing the period before and during isolation. In addition, about 75% of the records are broken-down vehicles and 25% accidents, in both periods analyzed. Both in the PP (Pre-isolation Period) and in the IP (Insulation Period), the broken-down passenger vehicles correspond to 52% of the

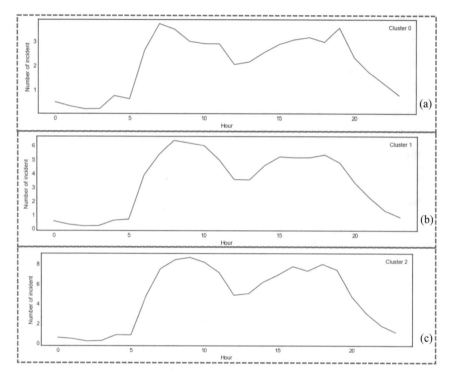

Fig. 3 Profile of attendance to incidents by cluster

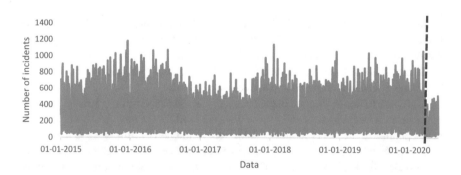

Fig. 4 Number of incidents attended per day

total attendance to broken-down vehicles and collision is the most frequent accident type. Therefore, from this analysis, it is noted that the profile of attended incidents do not change during the isolation period despite the reduction in the number of incidents.

Analyzing the data by vehicle type, Table 2 shows the difference in attendances in the periods under analysis (i.e., PP and IP). This table shows that, on average, 67

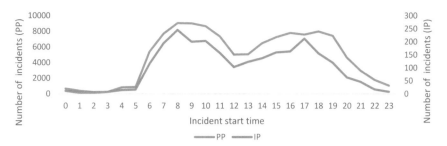

Fig. 5 Number of incidents per hour

passenger vehicles were serviced per day, before the isolation period, and that this average decreased to 34, during the pandemic time, which represents a reduction of approximately 53%. In relation to trucks, the average reduction was 46%. Reductions were also identified in less serviced vehicle types such as utility vehicles (25%) and motorcycles (37%).

To verify how significant this reduction is, a parametric hypothesis test is applied first, using the Z-Test, and, subsequently, a non-parametric test (Bootstrap Test), which compares the medians. Results are shown in Table 3. It is noteworthy that this method presents the results of the confidence interval calculated in three ways, using the Normal, Pivotal and Percentile Intervals methods.

To complement this analyses, Fig. 6 shows the service to passenger vehicles through a normal distribution. This example was used because this vehicle type corresponds to the majority of calls (75%). Thus, in Fig. 6, there is a change in calls between the two periods, but they are not completely different, which can be observed by the confidence intervals presented by Z-Test (Table 3). The normal distributions for the other vehicle type analyzed are like the one presented, thus the non-parametric hypothesis test is also used to assess whether the change in incident handling is really significant.

Taking into account the analysis of the medians by the Bootstrap Test, it becomes clear that the reduction in the number of incidents was significant, because when analyzing the intervals obtained by any of the three applied methods (normal, percentile intervals and pivotal), differences were observed between the data before and during the pandemic time. The reason is that, except for utility vehicles, all other confidence intervals exclude 0, thus demonstrating that really before the isolation the probability of an incident occurring was in fact greater than during isolation.

Finally, seeking to further improve the data analysis, another correlation is made on the relationship between traffic flow and the number of incidents, considering two of the most important roads in the Rio de Janeiro city which are: Avenida das Américas and Linha Vermelha. In these two roadways we can see an average reduction of 51% in the traffic flow in the IP. Figure 7 shows that in the days of lower traffic flows, the number of incidents also reduces.

Therefore, from the statistical analyzes employed in this study, it is possible to state that, despite the fact that there was no change in the profile of the incidence

Table 2 Comparison of the period before isolation with the period after isolation

Vehicle category	PP				IP			
	Mean	Standard deviation	Median	Confidence interval	Mean	Standard deviation	Median	Confidence interval
Car	67.074	17.849	66.5	31.374/102.773	36.416	11.891	34	12.633/60.199
Utility Vehicle	10.002	5.24	9	−0.478/20.482	7.52	3.622	8	0.275/14.766
Freight Vehicle	13.324	6.18	13	0.95/25.699	6.291	3.265	6	−0.239/12.823
Motorcycle	8	3.588	8	0.826/15.179	5	2.62	5	−0.259/10.259

Table 3 Parametric and non-parametric analysis

Vehicle type	Z test			Bootstrap test		
	Difference	Standard deviation	Confidence interval	Confidence interval		
				Normal	Percentile intervals	Pivotal
Car	30.657	21.448	−12.238/73.554	28.096/36.903	27.5/35.262	29.737/37.5
Utility Vehicle	2.481	6.37	−10.259/15.222	−0.634/2.634	0.5/3	−1/1.5
Freight Vehicle	7.033	6.996	−6.959/21.025	5.302/8.697	6/9	5/8
Motorcycle	3.003	4.448	−5.894/11.9	2.134/3.865	2/4	2/4

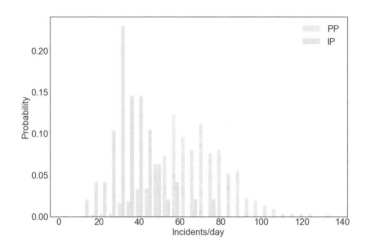

Fig. 6 Analysis for passenger vehicles

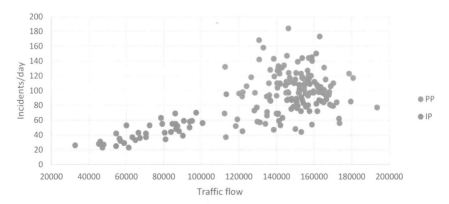

Fig. 7 Relationship between traffic flow and number of incidents

of incidents in the city, there was a significant reduction in the number of calls to traffic incidents, which can significantly impact in helping to reach several SDGs, as highlighted in Sect. 5.

5 Traffic Incidents and Sustainable Development Goals

As discussed in the previous sections, it is noted that the COVID-19 pandemic, although it caused serious economic and social losses, has resulted in a great reduction in the number of traffic incidents, which impacts, directly or indirectly, on nine of the 17 SDGs and 23 of the 169 targets of the 2030 Agenda for Sustainable Development [19], according to the exposed in Table 4.

With the reduction of incidents, there will be an improvement in road safety, due to the reduction in the number of traffic accidents. With 1.25 million people killed and up to 50 million people injured every year in road accidents, road safety must be a priority for the global community, not least because governments spend about 3% of their Gross Domestic Product on that [32]. Therefore, although there is no specific goal for road safety, it is included in two main references, which are: SDG 3 (Good health and well-being); and SDG 11 (Sustainable cities and communities).

The inclusion of traffic safety targets in the SDGs needs to mark a radical change in the world's response to road traffic injuries [33]. In this regard, in relation to SDG 3, the highly ambitious goal that strives to halve the global number of road fatalities and injuries by road traffic by 2020 (Target 3.6) represents a challenge for Member States in to reinvigorate its national road safety plans [34]. Regarding SDG 11, Target 11.2 stands out, which aims to provide access to safe, cheap, accessible and sustainable transport systems for all, improving road safety. In countries where traffic injuries are on a sustained downward path, this has been achieved by long-term political commitment [33].

It is also noteworthy that other SDGs may have an indirect relationship with the reduction of incidents, providing additional opportunities for synergy, such as: SDG 9 (Industry, innovation and infrastructure); SDG 12 (Responsible consumption and production); SDG 16 (Peace, justice and strong institutions); and SDG 17 (Partnerships for the goals). For example, road safety, in relation to SDG 9, has the ability to develop a quality, reliable, sustainable and resilient urban infrastructure (according to Target 9.1), in relation to SDG 12, it can encourage companies to adopt practices sustainable and integrate sustainability information in its reporting cycle (according to Target 12.6), in relation to SDG 16, it can guarantee a decision making by the responsive, inclusive, participatory and representative government (according to Target 16.7) and, in relation to SDG 17, it can increase the coherence of policies for sustainable development (according to Target 17.14), in order to mobilize partnerships with several stakeholders to solve this common difficulty. In this new era, intersectoral and innovative partnerships with various stakeholders will play a fundamental role in meeting many SDGs by the year 2030 [35], including reducing trade-offs and increasing synergy between them [36].

Table 4 SDGs and targets related to traffic incident

ID	Description
1	End poverty in all its forms everywhere
1.5	By 2030, build the resilience of the poor and those in vulnerable situations and reduce their exposure and vulnerability to climate-related extreme events and other economic, social and environmental shocks and disasters
3	Ensure healthy lives and promote well-being for all at all ages
3.4	By 2030, reduce by one third premature mortality from non-communicable diseases through prevention and treatment and promote mental health and well-being
3.6	By 2020, halve the number of global deaths and injuries from road traffic accidents
3.d	Strengthen the capacity of all countries, in particular developing countries, for early warning, risk reduction and management of national and global health risks
8	Promote sustained, inclusive and sustainable economic growth, full and productive employment and decent work for all
8.8	Protect labour rights and promote safe and secure working environments for all workers, including migrant workers, in particular women migrants, and those in precarious employment
9	Build resilient infrastructure, promote inclusive and sustainable industrialization and foster innovation
9.1	Develop quality, reliable, sustainable and resilient infrastructure, including regional and transborder infrastructure, to support economic development and human well-being, with a focus on affordable and equitable access for all
11	Make cities and human settlements inclusive, safe, resilient and sustainable
11.1	By 2030, ensure access for all to adequate, safe and affordable housing and basic services and upgrade slums
11.2	By 2030, provide access to safe, affordable, accessible and sustainable transport systems for all, improving road safety, notably by expanding public transport, with special attention to the needs of those in vulnerable situations, women, children, persons with disabilities and older persons
11.3	By 2030, enhance inclusive and sustainable urbanization and capacity for participatory, integrated and sustainable human settlement planning and management in all countries
11.6	By 2030, reduce the adverse per capita environmental impact of cities, including by paying special attention to air quality and municipal and other waste management
11.7	By 2030, provide universal access to safe, inclusive and accessible, green and public spaces, in particular for women and children, older persons and persons with disabilities
11.a	Support positive economic, social and environmental links between urban, per-urban and rural areas by strengthening national and regional development planning
12	Ensure sustainable consumption and production patterns
12.2	By 2030, achieve the sustainable management and efficient use of natural resources
12.6	Encourage companies, especially large and transnational companies, to adopt sustainable practices and to integrate sustainability information into their reporting cycle
12.7	Promote public procurement practices that are sustainable, in accordance with national policies and priorities

(continued)

Table 4 (continued)

ID	Description
12.8	By 2030, ensure that people everywhere have the relevant information and awareness for sustainable development and lifestyles in harmony with nature
13	Take urgent action to combat climate change and its impacts
13.2	Integrate climate change measures into national policies, strategies and planning
13.b	Promote mechanisms for raising capacity for effective climate change-related planning and management in least developed countries and small island developing States, including focusing on women, youth and local and marginalized communities
16	Promote peaceful and inclusive societies for sustainable development, provide access to justice for all and build effective, accountable and inclusive institutions at all levels
16.1	Significantly reduce all forms of violence and related death rates everywhere
16.7	Ensure responsive, inclusive, participatory and representative decision-making at all levels
17	Strengthen the means of implementation and revitalize the global partnership for sustainable development
17.7	Promote the development, transfer, dissemination and diffusion of environmentally sound technologies to developing countries on favourable terms, including on concessional and preferential terms, as mutually agreed
17.14	Enhance policy coherence for sustainable development
17.16	Enhance the global partnership for sustainable development, complemented by multi-stakeholder partnerships that mobilize and share knowledge, expertise, technology and financial resources, to support the achievement of the sustainable development goals in all countries, in particular developing countries

Source Ref. [19]

In addition, the reduction in the number of incidents influences the reduction of congestion and, consequently, reduction of Greenhouse Gas emissions and air pollutants [11]. The reason is that traffic incidents generate irregular but frequent road interruptions, which intensify congestion [37–40], increasing emissions in urban areas [14]. Thus, there is also an intrinsic relationship with SDGs related to climate change and extreme events such as SDG 1 (No poverty, more specifically Target 1.5 that seeks to reduce the exposure and vulnerability of the poor and those in situations vulnerable to extreme events related climate and other economic, social and environmental shocks and disasters), and SDG 13 (Climate action, with a focus on Target 13.2, which seeks to integrate climate change measures into national policies, strategies and planning). It should also be noted that the reduction of congestion also impacts the reduction of fuel consumption and carbon footprint, which is related to the SDG 12 (Responsible consumption and production through Target 12.2, aiming to achieve sustainable management and efficient use of resources natural).

6 Conclusions

This study sought to test the hypothesis that the isolation, due to COVID-19, resulted in a significant drop in the occurrence of traffic incidents, through the performance of statistical analyzes, and to relate this possible drop to the aid to the reach of SDGs and their respective targets. Therefore, from the statistical analyzes employed in this study, it is possible to state that, even though there was no change in the profile of the incidents in the city, there was a significant reduction in the number of calls, which can help, directly or indirectly, to achieve nine of the 17 SDGs and 23 of the 169 targets of the 2030 Agenda.

Thus, although the blockade carried out in various parts of the world to combat the spread of COVID-19 is not permanent, the findings found may represent the potential for mitigating traffic incidents in the future and assist in future incident management planning which has a great influence on meeting the SDGs and their targets of 2030 Agenda. In addition, the results show that collaborative efforts between researchers and public and private decision-makers will be necessary to gather and develop road safety strategies in relation to the new post-COVID-19 reality to promote a more sustainable future.

It is also noteworthy that as restrictions decrease, new patterns of behavior are likely to emerge. Thus, it will be important to monitor traffic incidents not only during the pandemic, but also in the post-pandemic period, as well as many of the ongoing social and behavioral changes that were initiated during the period of mobility restrictions. For example, social distance measures can encourage greater use of private transportation, after COVID-19, which can increase traffic incidents. However, work at home can become more prominent and decrease the need for daily commuting, which would reduce the number of vehicles circulating on the roads, and, consequently, reduce traffic incidents.

Acknowledgements This work was partially supported by the National Council for Scientific and Technological Development (CNPq), under grant #307835/2017-0. This work was supported by Carlos Chagas Filho Foundation for Research Support of the State of Rio de Janeiro, under grants #233926. This study was also financed in part by the Coordenação de Aperfeiçoamento de Pessoal de Nível Superior—Brasil (CAPES)—Finance Code 001.

References

1. De Vos J (2020) The effect of COVID-19 and subsequent social distancing on travel behavior. Trans Res Interdiscip Perspectives 5:100121. https://doi.org/10.1016/j.trip.2020.100121
2. Sutherland M, McKenney M, Elkbuli A (2020) Vehicle related injury patterns during the COVID-19 pandemic: what has changed? Am J Emerg Med 38(9):1710–1714. https://doi.org/10.1016/j.ajem.2020.06.006
3. Tosepu R, Gunawan J, Effendy SD, Ahmad AI, Lestari H, Bahar H, Asfian P (2020) Correlation between weather and Covid-19 pandemic in Jakarta Indonesia. Sci Total Environ 725:138436. https://doi.org/10.1016/j.scitotenv.2020.138436

4. Muhammad S, Long X, Salman M (2020) COVID-19 pandemic and environmental pollution: A blessing in disguise? Sci Total Environ 728:138820. https://doi.org/10.1016/j.scitotenv.2020. 138820

5. Loske D (2020) The impact of COVID-19 on transport volume and freight capacity dynamics: an empirical analysis in German food retail logistics. Trans Res Interdiscip Perspectives 6:100165. https://doi.org/10.1016/j.trip.2020.100165

6. Laverty AA, Millett C, Majeed A, Vamos EP (2020) COVID-19 presents opportunities and threats to transport and health. J R Soc Med 113(7):251–254. https://doi.org/10.1177/014107 6820938997

7. Dantas G, Siciliano B, França BB, da Silva CM, Arbilla G (2020) The impact of COVID-19 partial lockdown on the air quality of the city of Rio de Janeiro Brazil. Sci Total Environ 729:139085. https://doi.org/10.1016/j.scitotenv.2020.139085

8. Dutheil F, Baker SJ, Navel V (2020) COVID-19 as a factor influencing air pollution? Environ Pollut 263:114466. https://doi.org/10.1016/j.envpol.2020.114466

9. Venter ZS, Aunan K, Chowdhury S, Lelieveld J (2020) COVID-19 lockdowns cause global air pollution declines. Proc Natl Acad Sci 117(32):18984–18990. https://doi.org/10.1073/pnas. 2006853117

10. Federal Highway Administration—FHWA (2010) Traffic incident management handbook. https://ops.fhwa.dot.gov/eto_tim_pse/publications/timhandbook/

11. Baltar M, Abreu V, Ribeiro G, Bahiense L (2020) Multi-objective model for the problem of locating tows for incident servicing on expressways. TOP 29:58–77. https://doi.org/10.1007/ s11750-020-00567-w

12. Baltar MLB, Abreu VHS, Ribeiro GM, Santos AS (2021) Evaluating impacts of traffic incidents on CO_2 emissions in express roads. In: Muthu SS (ed) LCA based carbon footprint assessment. environmental footprints and eco-design of products and processes. Springer, Singapore. doi:https://doi.org/10.1007/978-981-33-4373-3_2

13. De Palma A, Lindsey R (2011) Traffic congestion pricing methodologies and technologies. Trans Res Part C Emerging Technol 19(6):1377–1399. https://doi.org/10.1016/j.trc.2011. 02.010

14. Grote M, Williams I, Preston J, Kemp S (2016) Including congestion effects in urban road traffic CO2 emissions modelling: do Local Government Authorities have the right options? Transp Res Part D: Transp Environ 43:95–106. https://doi.org/10.1016/j.trd.2015.12.010

15. Vingilis E, Beirness D, Boase P, Byrne P, Johnson J, Jonah B, Mann RE, Rapoport MJ, Seeley J, Wickens CM, Wiesenthal DL (2020) Coronavirus disease 2019: What could be the effects on Road safety? Accid Anal Prev 144:105687. https://doi.org/10.1016/j.aap.2020.105687

16. Lockwood M, Lahiri S, Babiceanu S (2020) Traffic trends and safety in a COVID-19 World. What is happening in Virginia? Virginia Department of Transportation (VDOT). ransportation Research Board (TRB) Webinar. http://www.trb.org/ElectronicSessions/Blurbs/180648.aspx.

17. Carter. (2020). Effects of COVID-19 Shutdown on Crashes and Travel in NC. North Carolina, Department of Transportation, Transportation Research Board (TRB) Webinar.

18. Shilling F, Waetjen D (2020) Special report (update): impact of COVID19 mitigation on numbers and costs of California traffic crashes. Road Ecology Center UC Davis. https://roadec ology.ucdavis.edu/files/content/projects/COVID_CHIPs_

19. United Nations—UN (2015) Transforming our world: the 2030 agenda for sustainable development - A/RES/70/1

20. Peeri NC, Shrestha N, Rahman MS, Zaki R, Tan Z, Bibi S, Baghbanzadeh M, Aghamohammadi N, Zhang W, Haque U (2020) The SARS, MERS and novel coronavirus (COVID-19) epidemics, the newest and biggest global health threats: what lessons have we learned? Int J Epidemiol 49(3):717–726. https://doi.org/10.1093/ije/dyaa033

21. Le Quéré C, Jackson RB, Jones MW, Smith AJ, Abernethy S, Andrew RM, De- AJ, Willis DR, Shan Y, Canadell JG, Friedlingstein P (2020) Temporary reduction in daily global CO2 emissions during the COVID-19 forced confinement. Nat Clim Chang 10(647):653. https:// doi.org/10.1038/s41558-020-0797-x

22. Cyberlab. (2020). Dados de contagem veicular. Available at: https://twitter.com/cyberlabsai?
    ref_src=twsrc%5Egoogle%7Ctwcamp%5Eserp%7Ctwgr%5Eauthor
23. Musselwhite C, Avineri E, Susilo Y (2020) Editorial JTH 16: the coronavirus disease COVID-
    19 and implications for transport and health. J Transp Health 16:100853. https://doi.org/10.
    1016/j.jth.2020.100853
24. Tirachini A, Cats O (2020) COVID-19 and public transportation: current assessment, prospects,
    and research needs. J Public Transp 22(1). https://doi.org/10.5038/2375-0901.22.1.1
25. Griffith SM, Huang WS, Lin CC, Chen YC, Chang KE, Lin TH, Wang SH, Lin NH (2020)
    Long-range air pollution transport in East Asia during the first week of the COVID-19 lockdown
    in China. Sci Total Environ 741:140214. https://doi.org/10.1016/j.scitotenv.2020.140214
26. Syakur MA, Khotimah BK, Rochman EMS, Satoto BD (2018) Integration K-means clustering
    method and elbow method for identification of the best customer profile cluster. IOP Conf Ser
    Mater Sci Eng 336:012017. https://doi.org/10.1088/1757-899x/336/1/012017
27. Bholowalia P, Kumar A (2014) EBK-means: a clustering technique based on Elbow method
    and K-means in WSN. Int J Comput Appl 105(9):17–24. https://doi.org/10.5120/18405-9674
28. World Health Organization—WHO (2020) WHO coronavirus disease (COVID-19) dashboard.
    https://covid19.who.int/
29. Ministry of Health [Ministério da Saúde – MS] (2020) Coronavirus panel [Painel Coronavírus].
    https://covid.saude.gov.br/
30. Brazilian Institute of Geography and Statistics [Instituto Brasileiro de Geografia e Estatística
    – IBGE]. (2020). Estimates of the resident population in Brazilian municipalities and federative
    units with reference date on July 1, 2020 [Estimativas da população residente nos municípios e
    para as unidades da federação brasileiros com data referência em 1º de julho de 2020]. https://
    biblioteca.ibge.gov.br/visualizacao/livros/liv101747.pdf
31. Tomtom (2019) Traffic index 2019. https://www.tomtom.com/en_gb/traffic-index/ranking/
32. World Health Organization—WHO (2015) Global status report on road safety. http://www.
    who.int/violence_injury_prevention/road_safety_status/2015/en/
33. Billingsley, S. (2016). Time for results: road safety and clean air for all, leaving no one behind.
    Brief for GSDR. https://sustainabledevelopment.un.org/content/documents/971129_Billin
    gsley_Time%20for%20Results--Road%20safety%20and%20clean%20air%20for%20all,%
    20leaving%20no%20one%20behind.pdf
34. Global NCAP (2020) Road safety and the united nation's global goals for sustainable
    development. www.stopthecrash.org/global-goals/
35. Fundación MAPFRE (2020) Road safety and the SDGs. https://noticias.mapfre.com/media/
    2020/01/Road-Safety-and-the-SDGs.pdf
36. Santos AS, de Abreu VHS, de Assis TF, Ribeiro SK, Ribeiro GM (2021) An overview on
    costs of shifting to sustainable road transport: a challenge for cities worldwide. In: Muthu SS
    (eds) Carbon footprint case studies. environmental footprints and eco-design of products and
    processes. Springer, Singapore. https://doi.org/10.1007/978-981-15-9577-6_4
37. Barth M, Boriboonsomsin K (2008) Real-world carbon dioxide impacts of traffic congestion.
    Trans Res Rec J Trans Res Board 2058(1):163–171. https://doi.org/10.3141/2058-20
38. Chung Y, Cho H, Choi K (2013) Impacts of freeway accidents on CO_2 emissions: a case study
    for Orange County, California, US. Transp Res Part D: Transp Environ 24:120–126. https://
    doi.org/10.1016/j.trd.2013.06.005
39. Sookun A, Boojhawon R, Rughooputh SDDV (2014) Assessing greenhouse gas and related
    air pollutant emissions from road traffic counts: a case study for Mauritius. Transp Res Part D:
    Transp Environ 32:35–47. https://doi.org/10.1016/j.trd.2014.06.005
40. Zhang K, Batterman S, Dion F (2011) Vehicle emissions in congestion: comparison of work
    zone, rush hour and free-flow conditions. Atmos Environ 45(11):1929–1939. https://doi.org/
    10.1016/j.atmosenv.2011.01.030
41. Rio de Janeiro's City Hall [Prefeitura do Rio de Janeiro] (2020) Rio Decree No. 47,282 of
    March 21, 2020 [Decreto Rio nº 47.282 de 21 de março de 2020]. http://www.rio.rj.gov.br/doc
    uments/8822216/11086083/DECRETO_47282_2020.pdf

Impact of COVID-19 on Sustainability in Textile & Clothing Sectors

Bhagyashri N. Annaldewar, Nilesh C. Jadhav, and Akshay C. Jadhav

Abstract There has been a severe setback towards the sustainability in textiles and clothing sector due to the outbreak of the COVID-19 global pandemic. There has been a loss of momentum towards sustainable textile production, and also the 3R's (Reduce, Recycle and Reuse) have taken a back seat. This brings into line the zero-waste hierarchy concept for a circular economy by taking specific initiatives like rethink/refuse/redesign, implementing 3R's, composting, material and chemical recovery, etc. The outbreak of COVID-19 has forced the textile sector to focus on developments that will enhance human beings' protection by producing antiviral clothing materials such as personal protective gears like private protective suits, masks, hand gloves, etc. This personal protective equipment (PPE) has generated a waste mountain of PPE worldwide and has created havoc in terms of sustainability. This chapter will emphasise the impact on sustainability in terms of textiles and clothing industries in detail. It will contribute towards cleaner production and sustainability in the textile sector because of COVID-19 by taking a step towards the opportunities for change in textile processes as per the laws. In addressing these inquiries, we use a structure for analysing sustainability and the textile and garments sector, featuring the chances and difficulties for a sustainable change concerning design, production, utilisation and end-of-life. This chapter's primary concern is to identify whether the ongoing pandemic will support the sustainability evolution concerning the textile and clothing sector.

Keywords COVID-19 · Sustainability · Textiles · Waste management · Supply chain

B. N. Annaldewar · N. C. Jadhav · A. C. Jadhav (✉)
Department of Fibres and Textile Processing Technology, Institute of Chemical Technology, University Under Section-3 of UGC Act 1956, Mumbai 400019, India

© The Author(s), under exclusive license to Springer Nature Singapore Pte Ltd. 2021
S. S. Muthu (ed.), *COVID-19*, Environmental Footprints and Eco-design of Products and Processes, https://doi.org/10.1007/978-981-16-3860-2_4

1 Introduction

Ever since the outbreak and development of COVID-19 in Wuhan city, China, during the late December of 2019, medical care units around the entire globe have battled to restrict the spread of the contamination. The causal specialist, SARS-COV-2, is a respiratory infection that belongs to the coronavirus family which is firmly identified with 2003 SARS-COV-1 known as severe acute respiratory syndrome coronavirus epidemic and 2013 MERS known as middle east respiratory syndrome outburst. The virus infection proliferates locally by human-to-human transmission due to hand to mouth transmission from defiled surfaces or due to inhalation of contaminated respiratory droplets are suspended in the air. The distinctive highlights of SARS-COV-2 are that it is profoundly infectious, with actual multiplication numbers going from 1.4 to 7.2 and is exceptionally pathogenic, with casualty rates in the scope of 1.4% in New Zealand to 14.9% in the United Kingdom. Within seven months since its first presence, COVID-19 has infected more than 18 million individuals over 180 nations and killed more than 700,000 individuals. The medical authorities urgently attempt to discover protected and successful medications to restrict illness seriousness and save lives.

At the hour of composing this article, an effective vaccination is not accessible. The medical authorities need to depend on conventional public health precaution techniques like testing, case recognition, contact tracing and isolation. These intercessions depend on why our wellbeing frameworks can rapidly and proficiently distinguish contaminated people and guarantee that they do not get into close contact with healthy individuals. Practically speaking, this is an incredible undertaking, as considerable extent of cases either do not show up any symptoms or have very mild symptoms. This is absolutely why government authorities around the world are suggesting general safety precautions on personal hygiene and sanitisation such as hand washing, social distancing and universal masking [1].

The COVID-19 pandemic has unfavourably hit the textile, apparel and fashion industries globally with colossal damage to organisations that cannot be measured now as the virus infection keeps spreading. The demand for textile materials worldwide and domestic markets has come to a halt because of the frenzy circumstance aroused due to the COVID-19 outbreak. Because of the lockdown, a wide range of textile industries are shut, and it is not easy to think about when those will be permitted to open. Labourers have been running to a great extent amid a wide range of disarray. The business sector is terrified because of money crunch, production network unsettling influence and labour-related issues. Stores are shut, and practically all buyers are dropping or delaying orders as they have extensive inventories. They may not place orders in the upcoming months too. The daily wage labourer who forms 80% of the labour force in textile and clothing manufacturing plants is on streets or back in his hometown terrified [2].

Developing consciousness of the textile business' ecological impression regarding chemical and water usage, overall carbon emissions generated and textile waste, natural resources utilisation, use of poisonous chemicals, generation of wastage, energy usage, carbon emissions and water contamination have concentrated on sustainability problems in textile and clothing sectors. Huge ecological effects occur during the numerous processes from fibre production, yarn preparation, cutting, sewing, weaving, knitting, dyeing, printing and finishing processes, and there is expanding burden on designers, brands and manufacturers to carry out feasible practices into activities at different processes as well as more extensively into essential administration and marketing approaches.

The research underlines the requirement for principal changes in the plan of action, including moderate fashion development and sustainable practices through the inventory network. Although the fashion industry business is broadly considered the most ecologically dangerous industry, it keeps developing, especially the short style fragment, which has a more prominent negative effect because of its emphasis on modest assembling, regular presentation of new products and transient product use. To neutralise these hurtful results, product and process advancements to implement sustainable manufacturing practices have arisen, for example, plan development with R's principle, closed-loop fashion framework and so forth. There is a more noteworthy need and criticalness to execute sustainable practices in textiles and garments sectors given their worldwide importance, size and significance [3]. Due to this pandemic, there is a danger that sustainability will drop off plans in the fashion business, as brands concentrate around endurance, intending to secure individuals, money and liquidity.

This chapter gives an overview of the impact of COVID-19 on sustainability in the textile industry. It also provides various solutions for sustainable development in the textile industry, which can help recover the sector from this pandemic.

2 Impact of COVID-19 on Textile and Clothing Industry

Among the most self-important monetary areas, the clothing business is one of the greatest helplessly influenced areas. The essential explanation can be the first contaminated country by COVID-19, China, known as the clothing business's primary source objective. The secondary cause can be the spread of COVID-19 in almost all the nations and regions worldwide, including the USA, Italy, Spain, Germany and so on and one worldwide movement. The global pandemic in COVID-19 has influenced the current activity methods of different textile sectors by posting limitation of social gathering, relocation of migrant workers and affecting every stakeholder directly from farmers to traders/exporters in the value chain textile sector. Few points are focussed on and referenced underneath:

- **Supply chain disruption**
  The exponential increase of COVID-19 cases throughout Asia, Europe and the USA has resulted in border closures and home quarantines. COVID-19 has posed a severe threat to the global supply chain because of the economic slowdown. The change in commodity consumption has disrupted supply, manufacturing,

logistics and sales. Hence, it has disrupted the global supply chain by weakening and slowing down international trade [4]. This effect of COVID-19 on the textile business has seen as a scourge emergency. The fundamental explanation is that the principal source of raw materials for the textile business, China, has been affected by the outbreak of COVID-19. Prompt after the lockdown announced in China, the entire inventory network of those enterprises relying on China got upset. The lead textile manufacturing nations' in the world are Bangladesh, Myanmar, Sri Lanka, Pakistan and so on are fundamentally dependent on China for their raw materials requirement. For example, Bangladesh is subject to China alone for more than 50% of clothing raw materials and around 40% of the machinery and spare parts for this industry.

Similarly, Myanmar is additionally reliant basically on China for around 90% of raw materials. As per an overview, it has been tracked down that 93% of Bangladesh suppliers announced that they confronted a deferral in raw material shipments during this pandemic. Other than these, because of these deferrals, the cost of the raw material has expanded. In Myanmar, it is anticipated that around 10% of plants in the Yangon region of Myanmar are now shut. In any event, 20 production lines across the entire Myanmar have been closed because of the lack of crude materials.

- **Cash flow constraints**

  The subsequent effect is order cancellation from the retailers and brands' end. The purchasers of this industry intend to delay future orders just like the current orders in handling. Another effect of COVID-19 on the textile business is the conceded instalment by the retailers. Because of the lockdown due to COVID-19, the deals of the retailers and brands go to nothing. Other than these, no transportation is accessible in this pandemic for the merchandise shipment, and other related undertakings likewise get stuck. Therefore, the brands and retailers informed the production line proprietors about the postponed payment instalment [5].

- **Unemployment**

  Coronavirus represents a genuine danger to worldwide general wellbeing in both developed and developing nations. The textile and clothing sectors' specialists have almost no instruction, are unskilled, have low financial and frequently provincial backgrounds and have less bartering power, bringing about their recognisable proof as vulnerable. The outcomes of the COVID-19 pandemic for these specialist workers are critical and incorporate vulnerability about whether they will be qualified for compensation during the COVID-19 pandemic and related issues, for example, absence of cash for essentials like food and worries about the re-opening of processing plants during COVID-19 disease. Other concerns are the health risks due to the lack of deterrent measures in the working environment and the improvement of psychological health conditions given the deficiency of employment and the dread of contracting COVID-19. With such shortcomings and in the lockdown conditions, starvation results from COVID-19 for the powerless and low-level pay. The COVID-19 lockdown and the resulting financial recession have prompted significant income loss for the poor working workers. They have driven away from home, break isolation rules and risk infection to this deadly disease trying to discover new ways to take care of their family [6].

- **Change in consumer behaviour**
  With additional time spent at home, consumers are being urged to reconsider their utilisation. Similarly, numerous individuals make garments purchases in front of special events, like weddings and vacation get-away; however, as a large number of these occasions have been cancelled or delayed, the impulse to purchase new garments is no longer there. Layoffs, furloughs and pay cuts are likewise influencing sales. Data from the industry propose the recurrence of fashions products purchased diminished during the beginning phases of the general health emergency. All the more significantly, buyers' psychological transmission capacity has also shifted from buying lifestyle needs, such as apparel, to day by day needs, such as food and beverages [7]. Likewise, customers prefer online-based shopping instead of visiting crowded shopping centres, which they casually did previously.

- **Shutting of retail stores**
  Retailers have reacted to the COVID-19 pandemic by covering their entryways. Numerous unmistakable clothing retailers worldwide, for example, H&M, Ralph Lauren, Chanel, Walmart, Sephora, Apple, Nike, Madewell, Everlane, Urban Outfitters and Lululemon have reacted to the Covid-19 pandemic by covering their entryways. Additionally, a couple of retailers, for example, Pink, Victoria's Secret and TJX are likewise shutting their online business websites incidentally. Significant shopping centres among all nations have additionally close [8–10].

- **Impact on Fashion events**
  COVID-19 hits the significant fashion events. Fashion weeks have been dropped in Melbourne, Beijing, Shanghai, Tokyo and Seoul. As the pandemic has advanced, creators need to apply their inventiveness to an assortment and how and where it will be delivered and exhibited. Designers have adjusted with producing and displaying their style products by streaming presentations online without a live audience present [11].

3 Use of Textile Materials for Avoiding the Spread of COVID-19

Protective gear comprises apparel put to shield medical services experts or any other person from the infectious disease. These, for the most part, include gloves, mask and gown. It will help to incorporate a face shield, goggles, surgical masks, gloves, surgical gown, headgear and safety boots for diseases that are spread by air or blood. Personal protective equipment (PPE) is essential to forestall COVID transmission in treatment centres and numerous activities like cleanliness, proper management of waste and internments and epidemic associated consideration. Scarcely any instances of personal protective equipment's are given underneath [12].

- **Surgical mask**

  A surgical mask comprises textile materials and plentifully utilised in offering protection to the patients infected with COVID-19. A surgical mask is manufactured using nonwoven fabric made up of polyester, viscose, polypropylene, etc., having a pore size of 0.3-10 μm. They are predominantly nonwoven; however, some of them are manufactured using cotton knitted fabrics. A surgical mask is obligatory in a crowded place where 6 foot of distances is complicated to maintain. This will help delay the transmission of the deadly coronavirus infection from individuals with no symptoms or from individuals who do not realise that they have been affected by the virus or not [13].

- **Surgical cap**

  A surgical cap is utilised in offering protection to the patients that are infected with COVID-19. The surgical cap is made up of polyester, polypropylene cotton, etc. They are additionally produced with cotton knitted textures. The surgical cap that goes with the surgical gown covers the medical specialist's head and also the hair, the ends. Surgical caps prevent any doctor or medical staff from splashing harmful fluids on the scalp. Nonwoven textile materials are broadly utilised for the production of surgical caps.

- **Surgical gloves**

  Surgical gloves are abundantly used to offer protection to the patients infected with COVID-19. Surgical gloves are made up of various mixed fibres in the form of a composite structure. Gloves are essential material to forestall, and they serve as protection in COVID-19 infected patient. Clinical gloves are manufactured with various polymers, nitrile rubber, latex, neoprene and polyvinyl chloride. These gloves are liberated with dust or powdered with corn starch to lubricate the gloves, making them simpler to wear.

- **Surgical gown**

  Surgical gowns are comprised of cotton, polyester, nylon, viscose or blended fabric. Surgical gowns are worn to prevent the doctor and the medical staff from getting infected from contagious diseases through vulnerable patients, like those with debilitated invulnerable immune systems.

- **Surgical shoe**

  A surgical shoe comprises textile materials like polyester, cotton, polythene or polypropylene, rubber, latex, etc. It assists with keeping medical staff foot clean and germ-free.

- **Ventilator bag**

  The ventilator bag is comprised of high thickness polyethene (HDPE) strands. In case a patient is attacked with COVID-19 and unable to breathe, ventilators are utilised to carry out the treatment of patients to encourage respiration [13–15].

4 Sustainability

Sustainability implies an ability to keep up some entity, result or process after some time. Sustainability can likewise be characterised as the practical and impartial dissemination of assets intra-generationally and between generationally with the activity of financial exercises inside the limits of a limited environment [16]. Sustainability includes the mix of ecological wellbeing, social value and monetary essentialness to make flourishing, sound, various and strong networks for this age and ages to come. Sustainability perceives how these issues are interconnected and require a frameworks approach and an affirmation of intricacy.

The most acknowledged definition of sustainability is the improvement that addresses the issues of the current age without compromising off the capacity of people in the future to addresses their problems [17]. A sustainable society needs to meet three conditions: its rates of utilisation of inexhaustible assets ought not to surpass their paces of recovery; its paces of the utilisation of non-sustainable assets ought not to reach the rate at which reasonable inexhaustible substitutes are created, and its paces of contamination of outflow ought not to exceed the assimilative limit of the environment [18].

5 Aspects of Sustainability: Environmental, Economic and Social

There are three aspects of sustainability which are discussed below:

- **Environmental**
  The environment can be characterised as the physical encompassing of man/woman. He/she is a part on which he/she depends on his/her exercises like physiological working, creation and utilisation—this actual climate environment from air, water and land to natural resources. Environmental degradation is an intense issue overall, which covers an assortment of issues, including contamination, biodiversity misfortune and animal extinction, deforestation and desertification, and global warming and much more. The environmental degradation and deterioration of the environment through the diminution of resources incorporate all the biotic and abiotic elements that structure our surroundings: water, air, soil, plant, animals and any other living and non-living component the earth. The primary consideration of environmental degradation is human generated (present-day urbanisation, industrialisation, overpopulation development, deforestation and so on) and natural calamities like (hurricanes, floods, rising temperatures, dry seasons, fires, etc.) [19, 20]
  Environmental pollution alludes to the degradation of the quality and amount of natural assets. Various types of human exercises are the fundamental reasons behind environmental degradation. For instance, the smoke transmitted by the

vehicles and processing industries extends the proportion of poisonous gases, which is recognisable worldwide. The waste things, smoke transmitted by vehicles are the central driver of pollution. Unconstrained urbanisation and industrialisation have caused air, water and sound pollution. Urbanisation and industrialisation help to grow the breakdown of the sources of water. So, addition, the smoke released by vehicles and substances like carbon monoxide, chlorofluorocarbon, nitrogen oxide and other clean elements causes air contamination [21, 22].

The harm caused by human beings to the environment is right now not included as an expense in monetary and social terms. This absence of "environmental value" has permitted us to over-misuse "free" natural resources—which are, obviously, not free. It has additionally prompted over-production of cheap products with exceptionally short life expectancies, which are generously disposed of into the environment after utilising, and afterwards, new affordable merchandise are bought and disposed of again. This cycle continues endlessly, influencing the planet's ability to re-establish its ecological administrations eventually [23].

Environmental sustainability is about the natural habitat and how it stays profitable and strong to help human existence. Environmental sustainability relates to biological system honesty and conveying limit of natural habitat. Environmental sustainability improves human wellbeing assistance by securing crude materials utilised for human necessities and guarantees that sinks for human squander are not surpassed to forestall damage to people. The ramifications are that natural resources should be collected no quicker than they can be recovered, while waste should be produced no quicker than the environment can acclimatise them. This is because the earth frameworks have limits inside which balance is maintained [16, 24].

- **Economic**
  The centre necessity of sustainability is that current financial exercises would not bring about an unreasonable burden on people in the future. This rule is sufficiently enough to suggest distinctive choice principles for protection. Sustainability advancement requires upkeep of the natural and human resource base fundamental for long haul financial development of ecological resources [25].

  Economic sustainability is utilised to characterise different techniques that promote financial resources for their best potential benefit. A sustainable economic model recommends a fine dispersion and professional designation of resources. The thought is to encourage the utilisation of those assets effectively and capably, giving long-term benefits and setting up productivity. The pleasant thing about adopting an all-out strategy for sustainability is that if you focus on social and ecological issues, benefits will frequently follow. Social activities affect consumer behaviour and worker performance, while environmental actions, for example, energy productivity and contamination mitigation can directly affect diminishing waste. Economic sustainability ensures that the business makes a benefit, yet additionally that business tasks do not make social or environmental issues that would hurt the organisation's long-term achievement [26].

- **Social**
  The social component of sustainability depends on how equivalence and understanding of the interdependence of individuals within the society are the basic requirements for an adequate quality of life, which is the principal objective of development [19]. The social aspect of sustainability refers to, in broad terms, public strategies that help social issues. These social problems relate to our prosperity and incorporate features like health care, housing, education, employment, etc. So forth, it guarantees that people do approach social administrations, do not endure lack of information on their privileges and exercise a dependable impact on advancing social strategies and amenities, both locally and nationally [20].

6   Sustainability in the Textile and Clothing Industry

The textile and clothing industries are the most polluting sectors on the planet. Its sustainability challenge includes different, organised and muddled issues [27]. Textiles and clothing currently play a critical part in the worldwide public dissertation on chemical society, environmental change, lack of water shortage and human liberties [28]. Their production and utilisation bring up a few issues and stresses that make difficulties over how individuals live their social, economic and political lives. Large numbers of the problems concern a few basic cultural and private practices and the job of multiples and frequently contradictory principles related to production and utilisation of textiles [29]. There are innovative solutions that settle some portion of the difficulties; others require dedicated activities concerning consumers, government, NGOs, business and others, and progressively so globally [30]. Especially organisations and customers have been distinguished as foremost entertainers now because of the quick fashion and textile industry's idea. Strategists and activists consider how organisations and consumers can be urged to assume liability, make voluntary strides towards sustainability upgrades and on occasion even be through and through compelled to change their decisions and practices [31]. The more profound inquiry is, notwithstanding, regardless of whether, how, and how much they can take on new liability for guaranteeing that the textile and clothing sector turns out to be more sustainable. Researchers stress on the problematic areas in the globalised textile, and clothing sector is profoundly intricate. Technological solutions aside, they find that adjusting behaviour and practice regularly requires the more complex task of working with changing the principles related to production and utilisation and being sensitive to various geographic, cultural and political contexts [32]. Aggregated studies likewise recommend that better data and correspondence from governments, activists, organisations, educational institutions, the media and others are vital [33]. Therefore, finding a proper solution for the various environmental, labourers' treatment and monetary issues inside the globalised fashion-driven textile and apparel market sector require innovative thinking and endeavours on both the supply and demand side of the market at different societal and administrative levels. The specific sustainability challenge is this requirement for an operative interaction between supply side and demand-side entertainers that brings feasible, sustainable principles and practices all the more straightforwardly into focus [34, 35].

7 Impact of COVID-19 on Environmental Sustainability in Textile Industry

The COVID-19 emergency has reemphasised the imperative role of plastic materials in our daily life. Plastics materials have contributed monstrously towards the medical services area and public health security throughout the ongoing global pandemic. Notwithstanding the inconvenience of countrywide lockdown, maintaining social distancing, limitation on voyaging and general get-together, continuous utilisation of hand sanitisers alongside utilisation of plastic-based personal protective equipment (PPE), viz. hand gloves, medical gowns, aprons, face safeguards, facial masks and other necessary PPEs for forefront health and safety officers and workers as prudent steps have also been implemented to maintain a strategic distance from infection defilement to battle the spread of COVID-19. Consumer's behavioural changes combined with the reliance on online shopping through e-commerce websites and take-away facilities for home-delivery of essential things during the global pandemic which have prompted an impressively expanded interest for plastic-based packaging materials, including single-use plastics contrary to the background of predominant boycotts or limitations in numerous nations. The contagion has also initiated a novel type of customer interest and social variations like frenzy purchasing, amassing food products and groceries among the majority and accordingly brought about an upsurge in the plastic-based packaging materials in numerous nations. A flood in the plastic demand throughout the worldwide pandemic is, subsequently, essentially because of PPEs and packing materials. Most of the PPEs are comprised of polymers like polyvinyl chloride (PVC), polypropylene (PP), low-density polyethene (LDPE), polyurethane (PU), polycarbonate (PC), while the plastics utilised in packing materials fundamentally comprises of polyethene terephthalate (PET), high-density polyethene (HDPE), low-density polyethene (LDPE), polystyrene (PS), etc.

Harmful and infectious COVID-19 biomedical waste (BMW) comprising of contaminated plastic-based PPEs and some other expendable things due to the affected sources like COVID-19 medical clinics, remote offices, containment zones, alongside comparative non-contaminated items from non-affected sources are created. In this manner, the COVID-19 biomedical waste generation can be straightforwardly connected to the great utilisation, for the most part, plastic-based personal protective equipment, and other non-reusable materials have been justified since the introduction of novel coronavirus epidemic. The increasing utilisation of plastic-based packing materials has combined with the expanding interest for medical items and packing materials amidst the pandemic has increased the plastic wastage around the world altogether. Accordingly, the pandemic has introduced a significant environmental challenge regarding plastic waste generation globally. Waste management amenities are commonly intended for consistent state activities with reasonable varieties in squander size and composition underneath ordinary situations. Nonetheless, the pandemic-incited alteration in waste creation and composition elements is almost certain to affect the current facilities' regular activity. Further, the decrease in plastic reusing due to plunging oil and petrol costs considering diminished transport activities in the crucial hour of pandemic-incited lockdowns has turned plastic waste management into a massive task [13].

Surgical masks and hand gloves ought not to be worn longer than a couple of hours and ought to be adequately disposed of to stay away from cross-infection. In this sense, a few nations have attempted to execute safety measures thinking about removing possibly contaminated PPE. For instance, the Portuguese Environmental Agency suggested that all conceivably contaminated PPE utilised by residents should be discarded as blended wastes (not recyclables) in fixed and sealed trash containers that will probably follow incineration facilities or will be subjected to landfilling. A few states in the USA have likewise quit recycling programme as administrations have been worried about the danger of COVID-19 spreading in recycling centres, consequently focussing on both incineration and landfilling. Such a decrease in squander recycling is different from the circular economy's objectives and sustainable development of events and surprisingly generating plastic waste pollution. By and large, PPE will probably wind up disposed of without precautionary steps alongside empty containers of hand sanitiser and organic wastes in regular municipal waste, or more regrettable, littered in the surrounding habitat. Erroneous removal of disposable gloves and face masks, alongside other plastic products, has been found littering in nearby public places [14].

There are various kinds of face masks accessible as indicated by their utilisation during the pandemic, like medical, filtering facepiece and non-medical, for example, cloth masks. All in all, medical face masks involve three layers, an external one of nonwoven fibres (they are usually resistant towards the water), a centre one is generally made up of a melt-blown filter, which is the essential filtering layer of the mask and an inward layer consists of soft fibres. Specifically, cloth masks are launderable and economical since they are produced with commercial synthetic materials like spandex, chiffon, flannel artificial silk, cotton quilts and among others. These engineered textile materials are made with polymers or polymers-normal fibre blends. The most utilised polymers in the assembling of these synthetic textile materials are polyester, nylon or polyether-polyurea copolymer. Subsequently, these kinds of textile materials may likewise add to the microplastics pool as fibres discarded during domestic washing into wastewaters and later reach wastewater treatment plants (WWTP). A laundry machine can trigger the subsequent discharge of these textile filaments into the seas. In April 2020, the utilisation of face masks for everybody out in the open spaces was obligatory in almost every part of the world. The manufacturing of face masks has expanded more than multiple times somewhat recently. This strong demand and importation and unnecessary use can prompt the fumble of medical waste by medical staff and residents because of the worldwide COVID-19 pandemic. Besides, the lack of information about the sort of domestic waste produced and its lack of arrangement by the people at home also contribute to the expanded plastic pollution during the pandemic. The present circumstance has cautioned scientific researchers because of the expansion of plastic waste in water bodies. Besides, cloth masks likewise address a peril to marine creatures since they can get tangled with the straps. In this sense, there are right now a few missions that request people to cut the straps from the face masks to evade animals getting caught in them.

Notwithstanding the mounting upsurge in PPE and dispensable plastics, the pandemic has produced substantial technological advances to dodge COVID infection. The utilisation of silver and copper nanoparticles with dynamic functionalities to battle microorganisms and guarantee asepsis illustrates this. South American nations, for example, Chile and Argentina, have showcased face masks with fungicidal, bactericidal, and antiviral properties and utilisation of spray and gels with copper nanoparticles. This innovation is being used to disinfect clinics, hospitals and nursing homes. It is identified that synthetic or engineered nanoparticles have been designated as arising contaminants. A few studies have reported nanoparticles' release in aquatic flora and fauna from commercial items and their drawn-out impact as a common pollutant in these water bodies and their extreme threat to marine animals [15, 36].

8   Solutions for Achieving Environmental Sustainability in the Textile Industry

8.1   Waste Reduction

Addressing and reducing PPE plastic contamination requires transdisciplinary cooperation among natural and social researchers, policymakers and waste managers on both municipal and national levels. In ensuring ourselves against COVID-19, we ought not to risk environmental health's integrity and make a future quandary as plastic pollution from PPE utilisation. To encourage legitimate disposal of the rapid and continuous increase in volumes of PPE waste from the society, we suggest the following measures [37]:

- Encourage general society to decrease the utilisation of dispensable gloves and wash hands frequently instead.
- Encourage and make available reusable face masks that can be now and again cleaned.
- Develop legitimate PPE disposal practices, find out about the subtleties of PPE waste management pathways and carry out improved collection techniques open to the general population.
- Raise far-reaching awareness on appropriate PPE discarding practices through focussed government promotion and training camps.

8.2 Minimise the Need for PPE in Health Care Settings [38]

The accompanying intercessions are recommended by World Health Organisation (WHO), limiting the utilisation and demand for PPE while guaranteeing to protect medical staff and others from the exposure to the COVID-19 infection in medical care settings is not compromised.

- Wherever possible, use telemedicine and phone hotlines to assess suspected cases of COVID-19 firstly, subsequently limiting the requirement for these people to go to medical care facilities for assessment.
- Use physical hindrances to lessen exposure to the COVID-19 infection, like glass or plastic windows. This methodology can be executed in territories of the medical care setting where patients will initially present, like emergency and screening zones, the patient enrolment desk area at the emergency division or at the pharmacy store window where the prescription is gathered.
- Postpone elective, non-urgent method and hospitalisations decrease the frequency of visits of severe patients with the goal that PPE can be rearranged to administrations in which COVID-19 patients get care.
- Cohort affirmed COVID-19 patients without coinfection with other contagious microorganisms in a similar room to smooth out the work process and encourage broadened PPE utilisation.
- Designate committed medical care workers/groups just for COVID-19 patient consideration so they can utilise PPE for longer timeframes.
- Restrict the quantity of healthcare workers from going into the rooms of COVID-19 patients on the off chance that they are not associated with giving immediate consideration. Smooth out the work process and diminish to a protected level consideration that requires face-to-face interaction between health specialist and patient.
- Consider utilising explicit PPE just if in direct close contact with the patient or when contacting the surrounding (for example, wearing a surgical mask and face safeguard, not utilising hand gloves or medical gown over the scrub suit, if going into the patient's room just to inquire or make visual checks).
- Visitors should not be permitted to visit confirmed COVID-19 patients, but if it is strictly necessary, confine the number of visitors and the time allowed; give clear instructions about why PPE is needed and why it should be utilised during the visit, about how to put on and remove PPE and perform hand cleanliness to guarantee that visitors avoid exposure.

8.3 Personal Protective Equipment (PPE) Reuse

COVID-19 brings massive difficulties in supporting the inventory supply chain for the single-utilised plastic PPE. Post-COVID-19, the transformations and adjustments in clinical practice will initiate enormous demand for PPE. Effective counter procedures for forestalling COVID-19 spread by incorporating alleviating probable high-risk aerosol transmission in medical care services utilising medical PPE and the suitable utilisation of face masks by the overall population that carries a lesser transmission jeopardy. PPE reuse is an expected temporary arrangement during the COVID-19 pandemic, where there is expanded proof for the effective positioning of reusing strategies. A few of the system are discussed underneath:

- **Dry and moist heat**
  Few researches have been carried out on the utilisation of various systems of heating for PPE processing. Heating usually causes irreparable physical harm in virus proteins that restricts binding to host cells. The issue with thermal methodology is to discard off COVID-19 with destructing PPE. It is very well be settled by decreasing contact time by up surging temperature. For instance, utilising 72 °C for 15 s gives a comparable degree of fatality to using a holding temperature of 60 °C for 30 min. The centres for disease control and prevention (CDC) expressed that, considering inadequate research available as of April 2020, moist heat has revealed a guaranteed probability to disinfect filtering facepiece respirators (FFRs).

- **UV irradiation**
  Ultraviolet (UV) irradiation causes the inactivation of many viruses by destroying the DNA or RNA through a photo-irradiation procedure. UV disinfecting exploits various wavelength bands where UVC (200–280 nm) is better than UVB (280–320 nm) and UVA (320–400 nm). Ideal irreversible molecular harm happens around the wavelength of 254 nm. There is a developing interest in the utilisation of UV innovations aimed at treating COVID-19 with other discoveries. The CDC also noticed that ultraviolet germicidal irradiation (UVGI) is a capable technique for PPE reuse, yet expressed that not all UV lights generate similar intensity and power; therefore, treatment times need to be changed likewise. Besides, UVGI is probably not going to deactivate all the viruses and other harmful organisms on an FFR because of shadow effects created via numerous FFR's construction layers [39].

- **Hydrogen peroxide**
  Hydrogen peroxide is a powerful oxidant and thrives as a disinfectant either as gas plasma, vapour or solution. For FFRs, hydrogen peroxide vapour (HPV) use is very much reported for not affecting filter performance and promising high throughput sterilisation. One method has been tried on FFRs inoculated with SARS-COV-2 and demonstrated to be effective, though the outcomes are only preliminary. It is right now being utilised in certain parts of the USA to sanitise enormous clusters of FFRs for redeployment to users with nitty–gritty depictions regarding how to carry out HPV use [40].

- **Ozone treatment**
  Another investigation reveals that ozone gas, a profoundly responsive chemical made from 3 oxygen atoms, might give a secure way to sanitising specific sorts of PPE that are popular for safeguarding medical care service personnel from

COVID-19. The team led by scientists at the Georgia Institute of Technology utilising two microorganisms like the novel coronavirus; further examination exhibited that ozone can deactivate viruses on things, for instance, polycarbonate face safeguards, Tyvek gowns, safety goggles and respirator masks without harming them as far as they do consist of stapled-on elastic straps. The investigation exhibited that the ozone treatment's uniformity and viability relied upon keeping up relative humidity of at least 50% in chambers utilised for sterilisation [41].

8.4   Use of Biopolymers for PPE

The entire solid waste management administration procedure of urban areas and countries has been affected by the remarkable amounts of expendable materials needed to protect against the virus infection. As one model, the aggregate sum of clinical waste created in Wuhan, where the pandemic began, is required to be nearly 25% more prominent in 2020 than in 2019. Supplanting the oil-based polymers used to make surgical outfits and masks with biopolymers would decrease not just the sea contamination generated by these all-durable single-use materials, yet also their overall carbon footprint [42]. Chances will emerge to address this challenge over flawlessly associating study and entrepreneurial environments, creating another line of conceivably usable bioplastic material.

A bioplastic comprised of plastic made somewhat or entirely from natural polymers produced from organic sources like sugarcane, potato starch, straw and cotton or the cellulose from wood. Some bioplastics decompose in the outer environment, while others are made with the goal that they should be composted in an industrial composting plant, supported by bacteria, enzymes and fungi. Bioplastics are generally made to be chemically indistinguishable from ordinary industrial plastics. Plastics produced using organic materials usually require a minimum amount of energy to have yet are similarly recyclable. They utilise fewer pollutants through the manufacturing process. Per ton of finished materials, the worldwide global warming effect on bioplastics' manufacturing process is often very considerably less than conventional plastics. Future green innovation exploration should be reached to form new biopolymer-based packaging and wrapping that include complex viruses and parasites [39].

9   Impact of COVID-19 on Social Sustainability in Textile Industry and Its Solution

The world of work is confronting a worldwide health emergency unlike any in the 100-year history. Simultaneously, this exceptionally globalised sector is battling with serious inventory side disturbance; as labourers are advised to remain at home, supply chains come to a standstill and production lines close. Notwithstanding the health risks presented by the virus infection, the industries' financial losses have influenced the business and livelihood of managers and labourers at the same time. Industrial facility and retail closures all around the world have compromised the feasibility of enterprises which led to labourers being suspended or losing their jobs. Small and medium-sized ventures (SME), a fundamental source of employment and development in the industry, will probably endure the most significant effect of this worldwide emergency.

Falling production and sales have had a massive thump on workers' impact, both regarding employment and working condition. For instance:

- An estimated 200 industries in Cambodia have either suspended or decreased production, and 5,000 labourers have lost their jobs positions.
- In Myanmar, a deficiency of raw materials from China has led to the closure of at least 20 factories and 10,000 jobs. At the same time, the number of orders has plummeted.
- In Vietnam, an expected 440,000 to 880,000 labourers could confront decreased hours or will become jobless. In the most-dire scenario, this figure could surge to as many as 1.3 million.
- In Bangladesh, as many as 2.17 million labourers have been affected by the global emergency, with many confronting unemployment as a maximum number of orders are cancelled, and production has been halted steeply. It is assessed that under 20% of firms can keep paying staff compensation for over 30 days under these conditions, and over 1,000,000 specialists have effectively been fired or furloughed. Non-payment of wages and the shutting down of processing plants is particularly hard for labourers in nations having very weak social protection frameworks [43].

Specialists concur that COVID-19 will additionally challenge working conditions and place labourers in a more dubious position. Critical impermanent and permanent job loss expanded utilisation of transient agreements, reduced wages and inconsistent payment will probably keep on affecting workers worldwide, which will give rise to an economic recession which will further lead to lower than pre-pandemic levels of interest in the near-term. For certain labourers, this will prompt delayed challenges in bearing the cost of essentials like house rent, food and other needful and mandatory supplies and may put a few workers in an undeniably unsafe circumstance both inside and outside of the work environment. These effects will probably be exacerbated by an absence of sufficient social assurance measures. As competition for jobs in the factories during the financial emergency will stay high, compliance with labour laws and codes of conduct may fall, and decent work shortfalls may rise. For instance, instances of union association busting, whereby labourers belonging to the associations are lopsidedly laid off contrasted with non-unionised labourers, have been reported in few nations. Moreover, decreased net revenues and precarious production requests resulting due to the pandemic may bring down compensation for labourers and increment surge orders; further, fuelling pressures connected to boisterous abuse attack in production lines, regardless of the expanded event of such practices, savagery and badgering just as gender equality will probably get less consideration in the close term because of COVID-19, as industry stakeholders will be engrossed with other pressing subjects such as occupational health and safety and have fewer assets to spend on projects to help labourers [44].

9.1 Solutions for Achieving Social Sustainability in the Textile Industry

COVID-19 is a pandemic which assaulted practically all nations of the world until now. Among all the financial sectors, the clothing business is perhaps the most affected sectors. The retail shops are shut with no income, which prompts the request to cancel the order to the clothing manufacturing suppliers. Accordingly, the workers have tumbled upon a depriving circumstance. They do not get their fair compensation, just like any help from any partners or stakeholders. They need to spend a solitary day with a tragic life. In this way, it is necessary to think for the workers at any rate since it is the subject of their endurance. The public authority can uphold the law that no production lines can be shut/laid off without satisfying the labourer's obligations. Indeed, if any crisis emerges for production line closure, the manufacturing plant needs to guarantee the labourers' wages until the circumstance recuperates, as no labourers can be terminated. The brands and consumers may help the workers by proceeding with the order and make the payment in due time. The monetary establishment may uphold the sector by giving credits and other financial backings. A multi-stakeholder initiative, including brands and retailers, bosses, governments, workers union and other related partners, is in an urgent emergency [45].

To battle against the emergency, at present, a few activities are taken by different corporations. For example, ten global associations' alliance has encouraged style brands and governments to secure a piece of garment labourers and future-verification supply chains in the middle of the COVID-19 global pandemic. To battle amplified disparities and secure textile labourers livelihood. Whose jobs are in question because of the COVID emergency, the alliance encouraged processing plants to "guarantee on-time payment of wages to labourers who remain effectively utilised. They have likewise asked employers that if offices need to close briefly, it ought to be a top concern of all stakeholders, all things considered, to help labourers straightforwardly or get funds to connect this period that they cannot work. At the point when worker reduction cannot be stayed away from because of long-term manufacturing plant shutdown or due to bankruptcy, all labourers ought to get their full legitimate privileges, including wages, advantages and severance pay. The alliance further desires to execute defensive measures in plants, for example, setting up healthcare facilities to limit the danger infection, including labourers in the decision-making process, giving them admittance to worker union and providing safe transportation [46].

10   Impact of COVID-19 on Economic Sustainability in Textile Industry and Its Solution

Still, there is no estimation about the financial damage that occurred because of COVID-19. The business analyst concurred that it would contrarily affect the worldwide economy. China's President Xi Jinping, speech on TV 23 February 2020 "It is unavoidable that the novel Coronavirus pandemic will significantly affect the economy and society". At the G20 congregation in Riyadh, Saudi Arabia, dated on 24 February 2020 Japanese Finance Minister Taro Aso articulated: "The spread of the new coronavirus is a public health emergency that could represent a genuine danger to the macro scale economy through the halt in production activities, interferences of individuals' development and cut-off of supply chains". The principal explanation behind it that China, known as "the global" factory, is the manufacturing centre for much of the worldwide business, i.e. energy, automobile, steel, textile, agribusiness, coal and electronic gadgets including mobiles, and so China has become one of the significant suppliers for the intermediate product to the final producers. China is represented 20% of the worldwide manufacturing intermediate products exclusively, but it was just 4% in 2002. Due to COVID-19, it is expected that worldwide development could be sunk by around 50% in 2020, contrasted with the assumption in November 2019. Other than these, the yearly worldwide GDP growth is assessed to reduce to 2.4% in 2020, all in all, which was 2.9% in 2019 [47].

Fashion is the second biggest market for consumer products after food and beverages. Also, the fashion business plays a focal part in creating income and employment, utilising more than 60 million individuals around the world. In certain countries, the sector is an essential contributor to national industrial yield and domestic value-added. The share of textile and clothing in manufacturing is high in many nations. The COVID-19 pandemic made a dramatic contraction in demand and production. The fashion business has been among those sectors all the more seriously hit by the coronavirus emergency [48]. The worldwide textile market was assessed to decline from $673.9 billion in 2019 to $655.2 billion in 2020 at a compound annual growth rate (CAGR) of—2.8%. The decline is principal because of a financial slowdown across several nations inferable from the COVID-19 outbreak and the measures to control it [49].

10.1   Solutions for Achieving Economic Sustainability in the Textile Industry

- **Managing Supply Chain Disruption**
  Given that the probable outbursts upset the input-sourcing, organisations should consider altering the sourcing mix to likely enhance hazard. When both suppliers

are found close to topographical proximity of each other, double sourcing methodologies are exposed to more noteworthy lockdown interruptions. Likewise, organisations with geologically various organisations of suppliers are as yet revealed to supply chain network interruption if a product depends on contributions from numerous suppliers as a solitary disturbance can have an ensuing consequential ripple effect [50].

The administration should figure out which of its products are especially exposed to solitary source dependencies or isolated area dependences and hope to assemble proper risk management methodologies. Temporarily, this could incorporate redistributing stock across districts or reduce reliance on products in danger of disruption. In the intermediate-term, organisations can hope to fabricate "buffers" to moderate the ripple effect only when a solitary supplier is undermined. This should be possible in two fundamental ways: (1) organisations can generate an inventory buffer, or 'safety stock of crucial components and products, and (2) organisations can generate a period buffer by deferring the manufacturing of products where demand is eccentric [51, 52].

Given the possibility of flare-ups disturbing input-sourcing, organisations should consider changing the sourcing blend to all the more likely expand hazard. Double sourcing methodologies, where the two providers are found close geological near each other, are presented to more major lockdown interruptions. Essentially, organisations with topographically various providers of providers are presented to inventory network disturbance if an item depends on contributions from multiple providers as a solitary interruption can have a resulting gradually expanding influence. The board should figure out which of its items are especially presented to single-source conditions or single area conditions and hope to construct fitting danger the executive's methodologies. For the time being, this could incorporate redistributing stock across districts or diminishing reliance on items at risk of interruption. In the intermediate-term, organisations can hope to fabricate "cradles" to alleviate the gradually expanding influence when a solitary provider is undermined. This should be possible in two primary manners: (1) organisations can make a stock cushion, or "security load" of fundamental parts and items, and (2) organisations can make period support by postponing the creation of merchandise where the request is inconsistent [53, 54].

- **Managing Demand Disruption**
  As referenced, the interest in fashion merchandise has observed a dramatic drop because of the effects of COVID-19. With individuals progressively telecommuting and going out less now and again, it tends to be accepted that the idea of the request for some fashion materials will change. The association's store network and the chain management include recharging product supplies as they are sold, instead of transportation of all stock to regions towards each fashion season [55].

  This has allowed the group a severe level of adaptability in reacting to request a change. Accordingly, fashion industries ought to consider restocking all through the fashion season in light of genuine interest rather than towards the beginning of the season based on anticipated interest. This technique yields more superior agility in reacting to potential interest interruptions [56].

It was discovered that organisations with a solid online presence could support sales when retail shops are shut. For instance, ASOS is an online retail store that announced cancelling less than 1% of its Spring/Summer 2020 intake. Online product contributions are simpler to refresh than actual ones. Thus, organisations with online retail stores can quickly change their item subscription to suit innovative demand conditions. In the long haul, all fashion firms should look to upgrade their online presence. Extravagance organisations should put a specific spotlight on the advancement of "the digital experience" given their objective market is acclimated with a high quality of administration in stores and expects a similar on the web. This should be possible inside by putting resources into restrictive online store facades and unions with trustworthy online retailers. Ongoing advancements in expanded reality innovation can likewise be adjusted to make digital showrooms and fitting rooms. Firms in this manner search for creative approaches to building up their online presence in the post-COVID-19 business environment to moderate demand interruptions by improving the customer experience [57–59].

11 Outlook of Textile and Apparel Industry Post Coronavirus

The worldwide outbreak of COVID-19 has created chaos in the fashion industry. The spread of the infection will undoubtedly have real ramifications, and organisations have begun feeling the heat with store shutdowns and vulnerability in orders. Since the pandemic is still in its developing stage, it is hard to foresee the full degree of the effect. Nonetheless, some fundamental changes can be anticipated due to this pandemic and how it can shape the business again [60].

- **Global Demand of Medical Textiles Increases**
  Sales of medical protective equipment, including surgical gowns, surgical masks, hand gloves and protective clothing, have bounced radically. The supply of these items is not able to keep up with the rising demand. The quick spread of the infection across the globe has sensitised individuals to hygiene and healthcare protection. The massive market for medical protective stuff like face masks, hand gloves and hygiene products, for example, wipes, is expected to increase and sustain even after the end of the COVID pandemic. This is a rewarding opportunity for the textile sector sooner rather than later [61].
- **Increased Focus on E-Commerce Sales and Digitalization of Supply Chain**
  To keep away from groups' occurrence and lessen the spread of coronavirus, shopping centres and retailers found a way to close their physical stores. The e-commerce websites of these stores are operational in specific nations. During the underlying period of COVID, shoppers expanded their online buying as a protected option in contrast to visiting actual stores. This move could prompt a changed

purchasing behaviour after the pandemic and fabricate long-time e-commerce business clients. Brands and retailers are additionally headed to consolidate digital strategy in their purchasing process. Online commercial websites are required to become more famous as brands and retailers hope to augment advanced choices of displaying their products and encouraging the purchasing and selling process [62].

- **Shorter textiles supply chain**
  The COVID-19 crisis can go about as an accelerator for choices for change that is long past due in the fashion sector and can at this point no longer be delayed. There are likewise various and interconnected drivers in the business environment that need significant changes in textile value chains to work on a worldwide and domestic level. The COVID-19 emergency has revealed insight into the dangers of depending solely on long worldwide supply chains. Closeness to consumer markets, nearshoring and "safe shoring" can be a business threat reduction technique. Simultaneously, it is also a methodology that can make a client-focussed supply chain network stronger to interruptions [62].

12   Conclusion

An informative examination dependent on the overall survey on the impact of COVID-19 has been presented in the chapter above to uncover the explanations to attain sustainability in the world's textile and clothing industry during this pandemic. The recent outburst of the coronavirus disease (COVID-19) has uncovered the textile and clothing industry's fragility. It is observed that the force predominance of apparel brands, unapproved subcontracting of clothing manufacturing units and utilisation of provisional workers by suppliers are the significant causes tormenting the public security aide in textile clothing supply chains working worldwide. As a relief strategy, suppliers, brands, industrial committees and NGOs should work as one after the COVID-19 pandemic to address the absence of government-managed security in the textile and clothing sectors inventory network. Another sourcing model, which would incorporate disturbance hazard sharing agreements and social security benefits for the labourers, should be approved. Likewise, brands ought to energise suppliers' utilisation of a long-lasting labour force by tuning the supplier choice and request for allocation strategy. This chapter's primary purpose was to identify the significant reasons behind the lack of sustainability caused due to the pandemic in the textile clothing supply chain operations worldwide. Even though this work chapter primarily focuses on the sustainability in textile industries during the COVID-19, its discoveries and consequences have a comprehensive impact on textile industries that mainly hinge on skilled labourers. Lastly, the textile, clothing and fashion manufacturing industries should implement strategic techniques and tactics to tackle similar situations shortly.

References

1. Beesoon S, Behary N, Perwuelz A (2020) Universal masking during COVID-19 pandemic: can textile engineering help public health? A narrative review of the evidence. Preventive Med 139:106236
2. https://www.fibre2fashion.com/industry-article/8635/covid-19-lockdown-impact-on-textile-industry
3. Islam MM, Perry P, Gill S (2020) Mapping environmentally sustainable practices in textiles, apparel and fashion industries: a systematic literature review. J Fashion Market Manage Int J 25(2):331–353
4. Chakraborty S, Biswas MC (2020) Impact of COVID-19 on the textile, apparel and fashion manufacturing industry supply chain: case study on a readymade garment manufacturing industry. J Supply Chain Manage Logistics Procurement 3(2):181–199
5. Sen S, Antara N, Sen S, Chowdhury S (2020) The unprecedented pandemic "COVID-19" effect on the apparel workers by shivering the apparel supply chain. J Text Apparel Technol Manage 11(3):1–20
6. Kabir H, Maple M, Usher K (2020) The impact of COVID-19 on Bangladeshi readymade garment (RMG) workers. J Pub Health 43(1):47–52
7. Brydges T, Retamal M, Hanlon M (2020) Will COVID-19 support the transition to more sustainable fashion industry? Sustain Sci Pract Policy 16(1):298–308
8. https://textilevaluechain.in/2020/04/01/impact-of-covid-19-coronavirus-on-global-and-domestic-market-of-textile-and-fashion-industry/
9. https://en.wikipedia.org/wiki/Impact_of_the_COVID19_pandemic_on_the_fashion_industry
10. https://www.tpci.in/indiabusinesstrade/blogs/tangled-web-the-impact-of-coronavirus-on-indian-textile-industry/
11. Aneja A, Pal R (2015) Textile sustainability: major frameworks and strategic solutions. In: Muthu SS (ed) Handbook of sustainable apparel production. CRC Press, Boca Raton, pp 289–306
12. Alam SMM, Islam S (2020) The implication of textile materials applied in preventing the spread of COVID-19. Biomed J Sci Tech Res 28(5):22009–22018
13. Parashar N, Hait S (2021) Plastics in the time of COVID-19 pandemic: protector or polluter? Sci Total Environ 759:144274
14. Silva ALP, Prata JC, Walker TR, Duarte AC, Ouyang W, Barceló D, Rocha-Santos T (2021) Increased plastic pollution due to COVID-19 pandemic: challenges and recommendations. Chem Eng J 405:126683
15. Ardusso M, Forero-López AD, Buzzi NS, Spetter CV, Fernández-Severini MD (2021) COVID-19 pandemic repercussions on plastic and antiviral polymeric textile causing pollution on beaches and coasts of South America. Sci Total Environ 763:144365
16. Mensah J, Casadevall SR (2019) Sustainable development: Meaning, history, principles, pillars, and implications for human action: a literature review. Cogent Soc Sci 5(1):1653531
17. Barooah N, Dedhia EM (2015) Study of socio-economic status of women engaged in handloom weaving and measures for enhancing their sustainability. Int J Res Soc Sci 5(4):653–665
18. Alhaddi H (2015) Triple bottom line and sustainability: a literature review. Bus Manage Stud 1(2):6–10
19. Mamidipudi A, Bijker W (2012) Mobilising discourses: handloom as sustainable socio-technology. Econ Polit Weekly 47(25):41–51
20. Maurya PK, Ahmad SAAA, Zhou Q, da Silva J, Castro EK, Ali H (2020) An introduction to environmental degradation: Causes, consequence and mitigation. Environmental degradation: causes and remediation strategies. Alteration in Haematological Indices of Heavy Metals Pollution in the Kali River, Uttar Pradesh, India, pp 1–20
21. Kumar V, Singh J, Kumar P (2020) Environmental degradation: Causes and Remediation Strategies. Vol. 1. Agro Environ Media, Publication Cell of AESA, Agriculture and Environmental Science Academy, pp 1–21

22. Chopra R (2016) Environmental degradation in India: causes and consequences. Int J Appl Environ Sci 11(6):1593–1601
23. Choudhary MP, Chauhan GS, Kushwah YK (2015) Environmental degradation: causes, impacts and mitigation. In: National Seminar on Recent Advancements in Protection of Environment and Its Management Issues (NSRAPEM-2015). Maharishi Arvind College of Engineering and Technology, Kota, India
24. Goodland R (1995) The concept of environmental sustainability. Annu Rev Ecol Syst 26(1):1–24
25. Foy G (1990) Economic sustainability and the preservation of environmental assets. Environ Manage 14(6):771–778
26. Balaji NC, Mani M (2014) Sustainability in traditional handlooms. Environ Eng Manage J 13(2):1–2
27. Boström M, Karlsson M (2013) Responsible procurement, complex product chains and the integration of vertical and horizontal governance. Environ Policy Gov 23(6):381–394
28. Boström M, Jönsson AM, Lockie S, Mol APL, P., O. (2015) Sustainable and responsible supply chain governance: challenges and opportunities. J Clean Prod 107:1–7
29. De Bakker F, Nijhof A (2002) Responsible chain management: a capability assessment framework. Bus Strateg Environ 11:63–75
30. De Brito MP, Carbone V, Meunier Blanquart C (2008) Towards a sustainable fashion retail supply chain in Europe: organisation and performance. Int J Prod Econ 114:534–553
31. Goworek H (2011) Social and environmental sustainability in the clothing industry: a case study of a fair-trade retailer. Social Responsib J 7(1):74–86
32. Harris F, Helen R, Dibb S (2016) Sustainable clothing: Challenges, barriers and interventions for encouraging more sustainable consumer behaviour. Int J Consum Stud 40(3):309–318
33. Laudal T (2010) An attempt to determine the CSR potential of the international clothing business. J Bus Ethics 96(1):63–77
34. Mol A (2015) Transparency and value chain sustainability. J Clean Prod 107:54–161
35. Niinimäki K, Hassi L (2011) Emerging design strategies in sustainable production and consumption of textiles and clothing. J Clean Prod 19:1876–1883
36. De-la-Torre GE, Aragaw TA (2021) What we need to know about PPE associated with the COVID-19 pandemic in the marine environment. Mar Pollut Bull 163:111879
37. Ammendolia J, Saturno J, Brooks AL, Jacobs S, Jambeck JR (2021) An emerging source of plastic pollution: the environmental presence of plastic personal protective equipment (PPE) debris related to COVID-19 in a metropolitan city. Environ Pollut 269:116160
38. World Health Organization (2020) Rational use of personal protective equipment for COVID-19 and considerations during severe shortages: interim guidance. No. WHO/2019-nCoV/IPC_PPE_use/2020.4. World Health Organization, Geneva
39. Rowan NJ, Laffey JG (2021) Unlocking the surge in demand for personal and protective equipment (PPE) and improvised face coverings arising from coronavirus disease (COVID-19) pandemic: implications for efficacy, reuse and sustainable waste management. Sci Total Environ 752:142259
40. Celina MC, Martinez E, Omana MA, Sanchez A, Wiemann D, Tezak M, Dargaville TR (2020) Extended use of face masks during the COVID-19 pandemic: thermal conditioning and spray-on surface disinfection. Polym Degrad Stability 179:109251
41. Toon J (2020) Ozone disinfection could safely allow reuse of personal protective equipment. Biomed Saf Stand 50(16):121–123
42. https://www.chemistryworld.com/features/sustainable-covid-19-protection/ 4012826.article
43. https://www.ilo.org/sector/Resources/publications/WCMS_741344/langen/index.htm
44. https://www.ilo.org/asia/publications/WCMS_755630/langen/index.htm
45. Sen S, Antara N, Sen S, Chowdhury S (2020) The apparel workers are in the highest vulnerability due to COVID-19: a study on the Bangladesh apparel industry. Asia Pacific J Multidiscip Res 8(3):1–7
46. https://fashionunited.uk/news/fashion/fashion-industry-coalition-fights-to-protect-garment-workers/2020042248606

47. Sen S, Antara N, Sen S, Chowdhury S (2020) The unprecedented pandemic COVID-19 effect on the Bangladesh apparel workers by shivering the apparel supply chain. J Text Apparel Technol and Manage 11(3):1–20
48. https://www.unido.org/stories/will-covid-19-accelerate-transition-sustainable-fashion-ind ustry#story-start
49. https://www.researchandmarkets.com/reports/5019783/textile-global-market-report-2020-30-covid-19
50. D'Adamo I, Lupi G (2021) Sustainability and resilience after COVID-19: A circular premium in the fashion industry. Sustainability 13:1861
51. Kumar A, Luthra S, Mangla SK, Kazançoğlu Y (2020) COVID-19 impact on sustainable production and operations management. Sustain Oper Comput 1:1–7
52. Akintayo WL (2020) Revitalising the nigerian textile industries for mitigating the effect of COVID-19 and achieving sustainable economic development. KIU J Human 5(3):57–66
53. Teng X, Chang BG, Wu KS (2021) The role of financial flexibility on enterprise sustainable development during the COVID-19 crisis—a consideration of tangible assets. Sustainability 13(3):1245
54. Wu HL, Huang J, Zhang CJ, He Z, Ming WK (2020) Facemask shortage and the novel coronavirus disease (COVID-19) outbreak: reflections on public health measures. EClinicalMedicine 21:100329
55. Barreiro-Gen M, Lozano R, Zafar A (2020) Changes in sustainability priorities in organisations due to the COVID-19 outbreak: averting environmental rebound effects on society. Sustainability 12(12):5031
56. McMaster M, Nettleton C, Tom C, Xu B, Cao C, Qiao P (2020) Risk management: rethinking fashion supply chain management for multinational corporations in light of the COVID-19 outbreak. J Risk Financ Manage 13(8):173
57. Taqi HM, Ahmed HN, Paul S, Garshasbi M, Ali SM, Kabir G, Paul SK (2020) Strategies to manage the impacts of the COVID-19 pandemic in the supply chain: implications for improving economic and social sustainability. Sustainability 12(22):9483
58. Yu H, Sun X, Solvang WD, Zhao X (2020) Reverse logistics network design for effective management of medical waste in epidemic outbreaks: insights from the coronavirus disease 2019 (COVID-19) outbreak in Wuhan (China). Int J Environ Res Public Health 17(5):1770
59. Sundari KM, Sowmiya MB (2020) Tirupur textile start-ups in COVID 19 pandemic situations-scot analysis. J Manag 2(7):81–88
60. de Abreu MCS, Ferreira FNH, Proença JF, Ceglia D (2020) Collaboration in achieving sustainable solutions in the textile industry. J Bus Ind Market
61. Impact of COVID-19 on the Indian Textile Industry. https://wazir.in/textile-apparel-insights/industry-report/
62. Mishra M, Mishra P (2021) Prioritising financial crises due to COVID-19: an economic safety and sustainability approach in India. Int J Syst Dyn Appl 10(1):1–11

Environmental Sustainability and COVID-19 Pandemic: An Overview Review on New Opportunities and Challenges

Bassazin Ayalew Mekonnen and **Tadele Assefa Aragaw**

Abstract COVID-19 pandemic rocketed the human-being anxiety with an exponential magnitude in the globe. The World Health Organization ruled wearing personal protected equipment (PPE), (both of the medical care professionals and the ordinary persons) to control the virus transmission. However, those PPE are linked to environmental damage due to ever-increasing activities as a root cause of pollutants in the soil, water, and air environment. The deterioration and damage of the natural environment by PPE pollutants contribution, the unsustainable has occurred, and the spread of the new wave of deadly pandemics. This review aimed to discussed the link between COVID-19 associated wastes and environmental impacts to be an unsustainable threat. The discussion included the opportunities and threats of the pandemic on natural environments such as air, water, and soil pollution level, especially air pollution reduction as an opportunity due to lockdown in the globe that minimizes the greenhouse gas (GHGs) emission, decreasing water pollution, and ecological restoration. Furthermore, the continued COVID-19 pandemic has driven extensive use of PPE around the world for virus protection augments biomedical wastes from healthcare hospitals, quarantine centers, and inhabitants resulted in the indiscriminate disposal threatens the environment with a new form of microplastics sources and litter pollution. In this regard, emerging and executing novel sustainable management strategies are paramount to reduce their impacts on the environment during and post-pandemic periods. In this way, this review highlighted the COVID-19 as an opportunity in terms of environmental sustainability toward the circular economy by adopting renewable energy sources and green practices as a sustainable option of waste management. Thus, the expected suitable implementation strategies were proposed in such a way that might be essential for environmental sustainability issues during the COVID-19 era.

B. A. Mekonnen (✉) · T. A. Aragaw
Faculty of Chemical and Food Engineering, Bahir Dar Institute of Technology, Bahir Dar University, Bahir Dar, Ethiopia

B. A. Mekonnen
Bahir Dar Energy Center, Bahir Dar Institute of Technology, Bahir Dar University, Bahir Dar, Ethiopia

© The Author(s), under exclusive license to Springer Nature Singapore Pte Ltd. 2021   117
S. S. Muthu (ed.), *COVID-19*, Environmental Footprints and Eco-design
of Products and Processes, https://doi.org/10.1007/978-981-16-3860-2_5

Keywords COVID-19 · Environment · Positive impacts · Negative impact · Sustainable management · Plastic pollution

Abreviations

NASA   National Aeronautics and Space Administration
BMW   Biomedical waste
ESA   European Space Agency
EEA   The European Environment Agency
NMVOC   Non-methane volatile organic matter.
IEA   International Energy Agency
GDP   Gross domestic product
SUPs   Sing-use plastics
PPE   Personal protective equipment
WHO   World Health Organization

1 Introduction

The current COVID-19 pandemic outbreak and spread have become globes series of health and environmental threats. The pandemic is set to have massive and sweeping effects on mankind and the entire planet as well. It postponed and unsettled modern society's activities in all circumstances [1]. The unusual outbreak of COVID-19 in the globe forced almost all nations under partial or total lockdown from a few weeks up to a few months. Accordingly, nations governors announce stay at home and confinement principle to sidestep public virus widespread [2]. Consequently, several human activities and public gathering events are canceled [3]. Besides, several manufacturing industries, road, and aviation transport services are banned from their regular working. Furthermore, the use of conventional energy sources has lessened considerably due to lower power requirements by manufacturing industries than usual [3]. As a result of the non-functioning of these services providing sectors effluents from them has cut largely result in nearly null GHG and aerosols emissions to the atmosphere.

Meanwhile, the struggles for prevention of SARS-CoV-2 spread, via confinement measure showed encouraging outdoor air quality improvement; the pandemic has set to destructive human economic activities, leading to worldwide economic destruction. For example, oil values reached negative while emissions have been noted lowest everywhere in the globe [4]. The current pandemic situation is bound to have significant implications on the three pillars the energy-environment-economy triangle. The consequence of this pandemic on the three pillars and sustainable growth approach is also heading for rather intricate and mixed. So, understanding the comprehensive crisis of COVID-19 and supper adjustment of the world to tackle upcoming challenges is vital for the future.

On contrary to worldwide economic and social destructions, COVID-19 imposed an opportunity and threats on the environment [3, 5]. Among the negative consequence, the COVID-19 preventive measures cause increased infectious medical and plastic pollution in the environment [5]. Huge single-uses-plastic production and use for preventive measures of the pandemic provoked global environmental and health crisis [6]. In addition, a drastic shift in life-saving in the pandemic is a clear indicator of plastic pollution and lesser attention to environmental issues [6]. As a result, excess production, use, and dumping of SUPs and PPE occurred these days [7]. Consequently, extensive public use has imposed cross-contamination on the environment due to improper disposal of SUPs and PPE debris with regular municipal solid wastes [5, 7]. In the long run, the disposal of PPE in open fields becomes a predominant source of microplastic in both aquatic and terrestrial environments [7–9]. Moreover, intense utilization patterns of SUPs, PPE, and behavioral shifts in waste management primacy are opposite to environmental sustainability [5].

Despite the pandemic has such negative consequences on the environment, the lockdown measure has an opportunity to improvements of the physical environment such as air, water, and soil [2, 10]. In this scenario, the reaction to the pandemic ban has an opportunity for environmental climate enhancement such as greenhouse gas pollutant species reduction and water quality improvements [11, 11]. In addition, ecosystems are being restored and dwellers are living through a clear sky for the first time. The place of natural beauty also experienced negligible visitor pollution levels. Due to the decrease in nitrogen oxide level, the ozone layer was found to improve to some extent [3, 13–16].

On the other hand, with the seedy upswing of confirmed virus cases, infected wastes from all levels of place and users' are being generated during the pandemic. For that reason, solid waste generation and controlling come to be complex and uninspiring [17]. Therefore, the pandemic outbreak has significantly augmented the volume of COVID-19-related medical waste in victim countries up to 18%–425% than the regular period [17]. In addition, the novel human coronavirus has also impacted MSW generation, type, and composition [17]. The pandemic has changed behavioral patterns and the lifestyle to extensive use of plastic use for assurance for virus survival. As a result, end-users shifted to SUPs and PPE preference for packaging and hygiene instead of perturbing plastic pollution [6, 17].

Currently, correlation amid COVID-19-crisis and the environment is a new incipient research focus and agenda. As a result, several investigations have been done on COVID-19 associated wastes as emerging environmental pollutants. This book chapter discussed the wide-ranging consequences of the COVID-19 pandemic crisis to the environmental sustainability issue and the future directives. Furthermore, it includes (1) the effects of COVID-19 on the natural environment; (2) the positive impacts of it on the greenhouse gas emissions reduction as an opportunity; (3) the

PPE wastes as a source of microplastic and macro-litter pollution during COVID-19 pandemics; and (4) the waste management regulations and implementation; and policy response.

2 Reviewing Strategies

A wide-ranging literature search on the current pandemic published articles (as research, review, baseline, short communication), podcast, news, and technical reports were retrieved from the direct Google search, Google Scholar, Scopus, and Web of Science databases, the web page including World Health Organization websites, news, and reports up to dates. The keyword searches included "COVID-19" and "the Environment," "novel coronaviruses" "Positive/negative impacts" and "COV-19," "environmental Sustainability" and "COVID-19." The downloaded documents were screened and examined relevant to the keywords and categorized based on research scopes of the present paper (the environmental crisis during the COVID-19; its impact and sustainable strategies during and post pandemics; and others).

3 COVID-19 Impacts on the Natural Environment

Beforehand of COVID-19, all the flora and fauna were fronting various ecological problems caused by pollution due to greenhouse gasses from manufacturing, transport, agricultural sectors, etc. The incidence of the pandemic has brought about opportunities and challenges on these surrounding issues. The new opportunities of the pandemic were manifested mainly by a decline in air smog, noise pollution, clean shorelines, a decline in the release of greenhouse gasses. The challenges imposed by the pandemic are a surge in plastic wastes at household and healthcare centers, a decline in waste recycling, indoor air pollution due to confinement, and many more.

3.1 Opportunities to Environmental Restoration

3.1.1 Outdoor Air Quality Improvement and GHGs Emission Reduction

Clean air is indispensable for human well-being. Its reported that about 91% of the world's inhabitants live in residences where poor air quality surpasses the allowable limits [18]. The penalties of air quality deficit are demonstrated in a substantial percentage of global death yearly [19]. Since 2016, the WHO report points out that air pollution accounted for about 8% of overall deaths in the world found in developing regions and some portion of Europe [18]. The emissions of several forms of carbon due to anthropogenic activities of human beings importantly degrade the air quality. However, the COVID-19 lockdown measures in almost all countries meaningfully enhanced air quality and reduced GHGs emissions. During the lockdown, manufacturing industries, transport sectors, and human movements have stopped resulted in a drastic decline in GHGs emissions [4].

The global modifications in pollutant intensities from restricted lockdown pollutants during the period February 2020 to June 2020 as a baseline compared with the 2019 pollution level are shown in Fig. 1. The analysis of Piers Forster et al. [20] data advocated that CO_2 emissions and total NOx emissions could have diminished by 30% in April 2020 driven by a reduction in surface-transport emissions. Particularly, emission reductions reached peaked in April 2020 where ~20% SO_2, ~5% CH4, ~7% N_2O, ~27% BC, ~14% OC, ~27% CO, ~20% non-methane volatile organic matter ~2.5% NH_3 reduction in global emissions April 2020 driven by a reduction [20]. The cooling around 0.01 ± 0.005 °C by 2030 and 0.3 °C by 2050 is likely to avoid future global warming due to the direct consequence of a pandemic-driven response.

The global reaction to the pandemic has led to an unexpected lessening of greenhouse gas releases and air contaminants mostly in all countries around the globe. For example, the lockdown measures a significant reduction (45–51%) of black carbon and NO_2 related to transportation emissions in the Barcelona span. In addition, the lower reduction PM_{10} (28–31.0%) and improvement in ozone levels (>50%) during lockdown was observed possibly due to lower titration of O_3 by NO and NOx decrement in a VOC-limited environment [21]. Similarly, an approximately 50% reduction of N_2O and CO happened because of the closure of heavy manufacturing industries

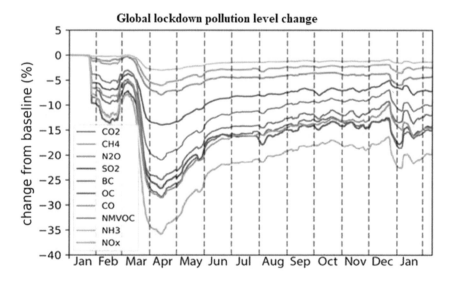

Fig. 1 The global pollution changes in different pollutants reduction levels from lockdown [20] adapted with permission of Nature, 2020

in China [22]. In the same way, China demonstrated a 20% cut of particulate matter ($PM_{2.5}$) in some towns [23].

Current data of NASA and the European Space Agency showed a 20–30% reduction in NO_2 in epicenters of COVID in Wuhan, Spain, Italy, France, and the USA [24]. Other reports of The European Environment Agency forecasted that NO_2 emissions fallen by 30–60% in several European cities in response to COVID-19 lockdown [25]. Likewise, NO_2 dropped by 25.5% in the pandemic period relative to earlier years of 2017–2019 levels in the USA [26]. Meanwhile, a reduction of the NO_2 level from 4.5 ppb to 1 ppb was found across Ontario, Canada [27]. Whereas, greenhouse gases reduction around 54.3% NO_2, 64.8% CO, 77.3% NO decline in concentration was observed in Sao Paulo-Brazil [28]. Correspondingly, reduction of NO_2 (35%) and CO (49%) was seen in Kazakhstan from the restriction measures of the pandemic[29], while a 50% and 62% discount of NO_2 level was stated in particular from the two big cities of Spain (Barcelona and Madrid), respectively, [30]. It was also reported a 5% cut of particulate matter (PM_{10}), 49% of SO_2, and 96% of NO_2 that were observed in Morocco[31]. In New Delhi, the capital of India the concentration of NO_2 and $PM_{2.5}$ was declined nearly by 70% during the domestic lockdown [32]. In addition, an overall reduction of $PM_{2.5}$ (46%) and PM_{10} (50%) was recorded in India during the imposed restrictions of the pandemic [33].

It is anticipated that transport sectors such as road and aviation are the main contributors of GHGs emissions (72% and 11%), respectively, [34]. The global movement restriction measure taken by nations impacts the aviation sector while several countries limited international and domestic travels. Due to this, the entry and departure of passengers in international and domestic flights were decreased and canceled by aviation sectors leads to deduction in GHGs emissions. For example, China cuts nearly 50–90% international and 70% local flights because of the pandemic related to January 20, 2020, which finally reduced approximately 17% of nationwide CO_2 emissions [35]. Furthermore, in a similar previous year, 96% of air flights were dropped worldwide due to the pandemic, which has eventual effects on the global economy and the environment[36].

The oil demand has declined by 2,850 litter universally in the head of three months of 2020, related to the similar period of the previous year[37]. Besides, the lesser energy appeal during the pandemic lockdown also reduced global coal consumption. For example, in India, coal-based power production was reduced by 26% with a 19% drop of entire power production later the lockdown [38]. China, the leading coal-based power user in the world, plummeted 36% compared to a similar period of the earlier year [38, 39]. Particularly, China reduced coal (50%) and oil (20–30%) use resulted in a 25% CO_2 decline, which is ~6% of global emissions [40, 41]. In this regard, an estimated cut of 1,600 Mt of CO_2, corresponding to over 4% of the global aggregate in 2019 could be found due to the pandemic crisis [41]. Likewise, a rough estimation of 17% reduction of daily CO_2 emission was observed similar to those in 2006 [42]. Generally, lower utilization of conventional fuels for transportation and manufacturing industries lessens burdens on the environment and the GHGs emission, which aids to fight climate change. However, the lockdown measure reduced indoor air quality due to stay-at-home confinment [43, 44].

3.1.2   A Decrease in Municipal Solid Wastes

The pandemic confinement underlines the need for households to apply segregation of household wastes in a separate collection. Consequently, residents gather MSW at home rather than at central collections leading to a decline in recycling. The announcement of the stay-at-home and travel stemmed resulted in a noticeable decline in waste generation in some areas of tourist destination cities and public places. For example, MSW in Macao, a usual tourist-target city in China, reduced by 17%–25% in the period February to May 2020 [45]. Similarly, the MSW generation in Morocco was decreased by 2%–10% from February to March 2020 compared to 2019 [46]. The lockdown measure in Italy led to a drop in the overall waste production by 27.5% in the city of Milan. The waste reduction during the pandemic comprises 24.4% leftover, 20% paper and cardboard, 16.7% glass, 16.3% plastic and metal, 14.4% household food and 80.5% commercial food waste in 10–14 weeks in the city [47]. In addition, the MSW generation in July 2020 at Trento, Italy, showed a reduction by 14% than in March 2019 [48].

The month of detention by the pandemic saw a 17% drop in MSW from 282.3 Mt to 242 Mt in Catalonia (Spain) [49]. Furthermore, the restricted mobility of visitors and commercial activities resulted in a 25% reduction in waste in Barcelona, accordingly, paper, glass, cardboard, organic, and packaging wastes were declined by 20%, whereas heterogeneous waste was declined by 12% [49]. Likewise, the volume of MSW getting junkyards cut by 20%–40% due to stopped recycling activities in India [50]. Similarly, the commercial waste volume was diminished by 57% in May 2020 in Tokyo, Japan [51]. Conversely, the pandemic has augmented waste generation in other regions. For instance, in England and Japan, it has been reported that MSW upsurges by 0%,20%, and 110%, respectively, higher than normal in May 2020 [51, 52]. Likewise, residential waste outgoing to landfilling increased by 34.7% in Tehran, Iran [53].

3.1.3   Water Quality Improvement and Pollution Reduction

The instant effect of the COVID-19 pandemic on water body systems and resources is inadequate, however, water quality and resources are perhaps affected on monthly and annual perspectives [10]. Due to the universal utilization of water for domestic and industrial applications, water pollution becomes unavoidable effluent dumped into rivers without treatment particularly in developing countries [54]. It is stated that water pollution load was reduced as the most important manufacturing sources of pollution have minimized or stopped in the lockdown period [54]. For example, the water quality of Vembanad Lake in India showed an improvement during the lockdown. According to the remote sensing study of the lake, an average 15.9% decrease in the suspended particulate matter was obtained during the lockdown period [54]. Also, the river Ganga and Yamuna in the same country have gotten a noteworthy level of pureness and met the permissible limit on the days of lockdown [55]. This improvement in the water quality of the lake was attributed to the sudden drop in invitees and 500% reduction of industrial sewages [50, 55]. Furthermore, the physic-ochemical parameter of the river Ganga meets the standard for domestic drinking water quality parameters during the restriction period in contrast to the pre-lockdown period [56–58]. Moreover, the restriction on public free movement activities were declined on water bodies in several places of the globe [2, 35, 59]. For instance, the Grand Canal found in Italy turned clear and several aquatic species have begun to reappear due to the ban of the pandemic [60]. Similarly, water contamination was also declined in the coastline areas of Asian countries [61, 62]. According to reports of Ref. [63], the COVID-19 lockdown reduced the quantity of food waste in Tunisia, which in turn reduces water and soil pollution. Specifically, textile factors utilized a huge quantity of water and this consumption is also reduced around the glove [64]. Besides, water and soil pollution caused by the huge quantity of solid wastes from construction and industries is also reduced. Moreover, the observed marine pollution in this pandemic era greatly reduced down to the global drop in import–export trade.

3.1.4   Noise Pollution Reduction

Noise pollution is the undesirable sound produced from anthropogenic human activi-ties [2, 65]. Usually, noise adversely affects the state of wellbeing of humans, together with circulatory disorders, high blood pressure, and sleep shortness [66]. It is reported that worldwide about 360 million citizens are susceptible to deafness caused by noise pollution [66]. According to WHO prediction, above 100 million citizens are exposed beyond the suggested limit of voice in Europe alone [67]. Moreover, environmental noise has hostile effects on wildlife and invertebrates shifting balance in predator and prey detection and the ecosystem [68]. The confinement and lockdown events ordered individuals to stay at home, which ultimately reduced noise level, economy, and mass communication in most cities worldwide [2]. It is stated that, in New Delhi, India, the noise level is reduced radically by about 40–50% in the contemporary lock-down period [50]. Similarly, the noise level of Govindpuri metro station (Delhi) is curtailed from 100 dB to 50–60 dB owing to vehicle restriction during the lockdown period [69]. In addition, the noise level of the residential area of the city is set to 40 dB and 30 dB in day and night time, respectively. As a result, city residents were a delight in the singing of birds in ranges from 40–50 dB [69]. Moreover, due to travel restraints, aviation flights and road transports have considerably reduced everywhere finally reduced the extent of noise pollution. For instance, in Germany passenger air flights were cut above 90%, car traffic was fallen by > 50% and trains were running at < 25% than the normal rates [70]. The COVID-19 lockdown reduced noise pollu-tion created by manufacturing industries, service providing industries and human activities worldwide resulted overall decline in the economic and social activities.

3.1.5   Ecological Repair and Clean Beaches

The tourism sector contributed considerable incredible growth to global GDP due to high-tech expansions and transport networks [71]. However, the tourism industry is responsible for an estimated 8% of global overall GHGs emissions [71]. In the nearby tourist attraction sites, plenty of hotels, restaurants, bars, and markets are built to facilitate and accommodate the visitors thereby demands lots of energy and other natural resources. For example, the carbon footprint of coastland hotel services of Spain consumed significant electricity and fuels have calculated maximum carbon emissions [72]. Due to this, natural beauty is usually experienced as a huge harsh [73]. Moreover, the visitors' rubbish dumps damage natural attractiveness and produce ecological disparity [74]. Restriction measures in the course of this pandemic have remarkably reduced tourist movement, thus retreating pressure on wild fauna and tourist spots around the world [2, 75]. Therefore, wildlife has been returning to inhabited areas from the previously fled due to the human confinement at home [75]. For instance, the clean beaches of Acapulco, Barcelona, and Salinas appear with crystal clear waters in Mexico, Spain, and Ecuador, respectively, [2]. Similarly, the restriction imposed by local government of Thailand on public gatherings and popular tourist arrival sites such as Cox's Bazar sea beach and Phuket showed an alteration in the color of seawater [59, 76]. Additionally, social distancing has a prominent amendment in the appearance of several beaches has been witnessed in the world. Consequently, the natural environment gets time to adjust to human displeasure and pollution reduction. For example, aquatic animals like dolphins recently returned to their previous inhabitant specifically on the coast of the Bay of Bengal, Bangladesh, and canals, waterways, and ports of Venice (Italy) after a prolonged decade [61, 62]. On the other way, many scientists have insisted nations for ruling out of wildlife trade permanently and encouraging afforestation for natural ecology restoration and prevention of future pandemic outbreak [3, 59].

3.2   Negative Impacts of COVID-19 on the Natural Environment

3.2.1   Increase in Biomedical Waste Generation

Global nations and cities that encountered high COVID-19 prevalence are striving to cope with the intense surge in medical waste generation for healthcare services [5]. Accordingly, massive medical waste is intensified universally and has become a risk to health and the environment during the pandemic era [2, 50]. Evidence showed that the daily output of COVID-19 medical waste was estimated 200 t/d to above 29,000 t/d by February 22 to the end of September 2020 throughout the world [17]. For example, in Jordan, the King Abdullah University Hospital generated(~650 kg per day) which accounted for ten times higher medical waste than the average generation rate of the usual working day of the hospital [77]. It was detected that a sudden surge in biomedical waste by 350% and 370% was also reported in Spain and China, respectively, [78]. Particularly, above 240 tons of PPE wastes were generated daily during the outbreak in Wuhan, China [13], which is 190 tons greater than the regular time [2]. The correspondingly central city of China generated ~247 tons of biomedical waste per day at the peak of the pandemic which is roughly fourfold greater than the pre-pandemic era [17]. Similarly, around 295 tons of COVID-19 PPE waste was engendered from early February 2020 to early March 2020 and projected over 2600 tone by mid-July 2020 in South Korea [79]. Whereas, generation of 80–110 tons per day of medical waste was recorded in Tehran, Iran, which causes (18%–62%) escalation of medical waste during the pandemic compared to the pre-pandemic [53]. Again, the quantity of biomedical waste is augmented by 400–450 kg/day at the start of lockdown in Ahmedabad city, India [50]. Similarly, ~206 Mt/day of medical waste was produced in Dhaka, Bangladesh [62]. Other Asian cities like Manila, Kuala Lumpur, Hanoi, and Bangkok experienced comparable rises, generating 154–280 Mt/day extra medical waste than the pre-pandemic period [80]. The US alone generated over 10,000 tons on September 30, far beyond other nations' medical waste generated in India (1690 tons), Brazil (2009 tons), and France (3776) [17]. Such unexpected escalation of precarious waste and their accurate management is a noteworthy task to the local authorities.

3.2.2   Source of Plastic and Microplastics Pollution

PPE is renowned for reliable and affordable preventive virus infection and transmission for the pandemic. Since the outbreak of COVID-19, people are exhaustingly utilized PPE for protection against viral infection. Hence, the massive production and dependency of plastic-based PPE are escalated in consort with waste worldwide [81]. Millions of PPE are being manufactured and used daily by ordinary citizens and front-line healthcare workers in the current pandemic. For instance, 129 billion face masks and 65 billion medical gloves are monthly used worldwide [82, 83]. Moreover, individual online shopping choices during restriction are also growing plastic demand. For example, packaged takeaway foodstuffs contributed an extra 1400 tons of plastic waste in a 2-month lockdown in Singapore [84]. As a result, the worldwide plastic packaging marketplace is estimated to grow by a 5.5% rate and $103.4 billion in 2021 higher than the year 2029 due to pandemic response. Overall, the demand for plastics is projected to rise by 40% over panic shopping for food packaging and 17% in other uses including medical purposes [82]. Meanwhile, SUPs trashes have been increased and observed with domestic and household wastes in the USA [62, 85]. In the same way, Wuhan Hospitals produced above 240 tons of SUPs biomedical waste daily, which six folds at peak of the pandemic than the average daily disposal before the pandemic incidence [86]. Similarly in the UK, it is foreseen that each inhabitant used one mask daily to produce a minimum of 60, 000 tons of polluted plastic waste [87]. Peace of studies showed that SUP was already among the foremost source of marine litter and mandatory PPE contributed the major share of the plastic liter [88].

For example, De-la-Torre et al. reported that 138 PPE items were recorded in 11 beaches during 12 sampling weeks along the coast of Lima, Peru [89].

Plastic pollution was mounting in the global ecosystem earlier to the COVID-19 pandemic [88]. Approximately, 4.8–12.7 million Mt of mishandled land-based plastic waste entered the aquatic system in 2010 only [90]. Among this quantity of waste 1.2–2.4 million Mt is transported by rivers [91]. The previous investigation stated that above 5 trillion plastic debris was floating in the oceans [92]. Furthermore, the imposed quarantine rules and panic online shopping have also led to a proliferation of packaged plastics in household wastes [2, 50]. Ultimately, such mismanaged SUPs and PPE will likely degrade into smaller pieces and become a source of microplastic pollution [82]. Recent explorations proved that PPE equipment mostly face masks are the likely source of microplastic fibers in the environment [7–9]. Additionally, such random disposal of plastic refuses clogged water-gate and degrades the environment [2, 81]. For instance, in the Magdalena River, Columbia, the ruin of nonwoven synthetic fabrics was the main source of microplastic fibers in water and sediment samples [82]. In addition, PPE and plastic littered in open environments will possibly persuade sewage system blockage. The occurrence and ubiquity of such plastic debris, allied associated contaminant triggered harmful effects on land-living and marine biota at different stages of biological organization [93–96]. Furthermore, plastic litter can be the vector for several pollutants, and pathogens such as the SARS-CoV-2 virus [80, 81]. Although recycling is effective for pollution averting, energy-saving, and natural resources conservation, it has been postponed in many countries due to the pandemic [97]. For example, the USA limited recycling plans by 46% in almost all cities owing to the frustration of the risk of COVID-19 distribution through recycling services [50]. Furthermore, the competitiveness of plastic recycling for fuel purposes had declined despite low oil prices and demand during the pandemic. Overall, due to interruption of regular recovery and recycling activities of MSW end up the landfilling of plastic pollutants in the global. Generally, the COVID-19 pandemic aggravated the difficulties of plastic waste management and hinder the efforts to cut plastic pollution [98]. Therefore, SUPs and PPE during the pandemic added extra plastic polymer to the existing plastic litter and endure the "never-lasting-story" of plastics in the environment resulted in the emergent research priorities [99] as shown in Fig. 2.

3.2.3 Miscellaneous Negative Impacts

Recently, a massive quantity of sanitizers is practical in public areas and individual level to sterilize pandemic viruses. Such widespread practice of decontaminators perhaps destroys non-targeted species, which probably exerts ecological imbalance. Moreover, the SARS-CoV-2 virus was detected in patient's feces and public wastewater of several victim countries [100–102]. So, extra actions in wastewater treatment are vital before discharge effluents municipal wastewater to the adjacent aquatic water bodies. For example, China enhanced the use of chlorine for disinfection to prevent

Fig. 2 Repercussions of COVID–19 on Plastic waste generation [98] adapted with permission of Elsevier, 2021

the current pandemic virus dispersal via wastewater. However, the undue application of chlorine in water could produce dangerous by-products [2]. In addition, the face masks disposed of on the earth cause entanglement and death to wildlife. For example, in Colombia, a bird was matted in a littered COVID-19 mask in a tree and died later it enfolded to its body and beak [103]. Additionally, the littered face mask is probably ingested as food mistakenly by invertebrates fill their stomachs and decline in food consumption leading to hunger and death [103].

The PPE trashes also transported to the aquatic ecosystem thereby creates plastic pollutions for aqua life and birds owing to ingestion. Moreover, marine plastic litter adsorbs contaminants which assist in the binding of the pollutant particles to the surface of the plastic [104]. Accordingly, it likely poisons the aquatic faunas through the ingestion of plastic and entanglement. In the end, the ingested plastic litter perhaps destroys or weakens faunas directly, the adaptation of them more susceptible to extra intimidations [105]. In addition, consumed plastic inhibits with impairment breeding and growth of offspring [78, 106]. Further, the disintegration of the macroplastic in the mask creates microplastics [106]. Hence, bioaccumulation of such microplastic arises in the food web to human-being then after causes accumulation of poisons [106].

4   Sustainable Strategies for the Environment Pollution Reduction During and Post-Pandemic

The extensive use and indiscriminate disposal of PPE by ordinary citizens became a debating issue between scarcity and correct handling in healthcare services while such utility is compulsory and of utmost importance for life savings. PPE disposal without decontamination and mixed with regular MSW worsen the littered in the environment. To tackle this problem, diverse handling and management methods have been implemented using incineration and landfilling [98]. Yet, these are not ideal preferences to promote the circular economy. Moreover, plastic ban or reduction is impossible in this pandemic era as citizens reliance on it for safety and hygienic purpose. In addition, an urgent adaption of the medical waste handling approach is required for the novel virus to spread to the surroundings. Therefore, examining for alternative routes to cope with plastic waste is indispensable.

For instance, the World Health Organization recommends black pins should be used for the separation of used PPE from general household garbage planned for recycling as shown in Fig. 3. Unlike the WHO, ABES Brazil, however, recommends disposing of used PPE in tightly tied plastic bags with domestic waste. In this sense, several nations tried to apply precautionary measures of handling and discarding PPE. As an example, the Portuguese Environmental Agency suggested every citizen disposed of PPE mixed wastes in wrapped bins rather than recycling that will preferably go to incineration and/or landfilling [107].

Similarly, U.S.A has prioritized incineration and landfilling and also stopped recycling strategies amid the threat of COVID-19 infection [2]. Such a suggestion on PPE waste handling and restriction on recycling contradict the traditional linear economy [109]. Although researchers attempt to manage the PPE and SUPs waste to value-added products, it ends up mixed with MSW to date, particularly in developing countries. Thus, the pandemic forced the continued improper use and handling of medical and plastic waste by billions of inhabitants with low biodegradation rates. Thus, urgent integrated response to waste management and treatments during and post-pandemic is paramount now and in the future. So, a guaranteed kind of waste management platform should be a part of the immediate reaction to a pandemic crisis. As huge plastic pollution in the pandemic resulted from: (1) greater demand on SUPs; (2) greater demand of PPE; (3) intensified medical waste; and (4) prioritizing of incineration and landfill. The following suggestions are proposed for the need future environmental suitability:

4.1   Using Biodegradable and Reusable Plastic PPE

To wipe out, the dependency of PPE and SUPs formulating reusing options to reduce the pressure of huge medical plastic wastes is required environmental care. Some of the ways out to reduce PPE waste could be rethinking the biodegradable and

Fig. 3 Correct disposal of PPE in the normal general/black bin waste adopted from NREB20-02 [108]

reusable PPE. Biodegradable PPE is one of the contemporary viable alternatives to cut the induced plastic waste. Therefore, plastic polymer derived from biological substances can be an alternative to substitute the existing polypropylene and PVC polymer of conventional plastic [110]. Some pieces of evidence have also shown that raw and waste natural fibers are possibly used to produce a cost-effective green bioplastic mask. For example, Sugar cane waste masks [111], Coffee-Based Face Masks [112], and Hemp Fiber masks [113] are existing green masks fabricated with 99.99% filtration capacity at this time [110]. In addition, these decomposable plastic decreases the CO_2 emissions by 30%–70% in comparison with conventional petroleum made plastics [114]. So, rethinking the current PPE constituting material substitution by low-carbon reusable options should be sought when possible.

Furthermore, biodegradable packaging foodstuffs or reusable fabric bags should be preferred over paper bags with no fees or bans in place [115]. Also, replacements of the existing plastic in the market with reusable, recyclable, or compostable ones likely reduced the SUP packaging and PPE [115, 116]. Moreover, incentivizing and restriction on fuel-based SUPs should continue by stakeholders. For example, fifteen Euro cents extra fees for each bag in 2002 led to a discount of 90% use of SUP bags in Ireland [88]. Likewise, reusable PPE should be practiced in pandemics providing they do not compromise public health.

4.2 Disinfection of Biomedical Wastes for Safe Disposal

All COVID-19 waste derives under the precarious BMW. Appropriate handling of the novel SARS-CoV-2 pandemic waste is boundless global fear to health and environmental sustainability. Thus, effectively managing through the appropriate disinfects and disposal methods are indispensable to regulate the pandemic spread and the environmental threat [117]. Ilyas et al. anticipated decontamination and reprocessing techniques of the COVID-19 medical wastes [117]. Among these treatment techniques, chemical (chlorine), vH_2O_2, UV radiation, incineration, and thermal treatment can effectively decontaminate and reduce the volume of the MBW for guaranteed recycling. The chemical decontamination with a 1% sodium hypochlorite solution is one of the best in-situ practices and easy to spray. In addition, this technique is not only to COVID-19 waste but is also effective to disinfect larger spaces, such as malls, hospital wards, and quarantine centers. Moreover, microwave-assisted sterilization practice is also suitable to cleanse PPE and cloths that can be recycled and reused. Equally, incineration is also a worthwhile process to disinfect a huge volume of COVID-19 waste despite energy-intensive operating at high temperatures (800–1200 °C) [117]. A strategy like "Identify, isolate, disinfect, and safe treatment practices" is effective for safer management of pandemic waste and the environment. Furthermore, following the CPCB Guidelines directive for the handling of pandemic waste is also vital for environmental protection and vires spread [118].

4.3 Redesign of Plastics from Fuel-Based Source and Repurposing It Later Use

The substitution of virgin plastics by fuel-based top-up raw materials and energy is previously being prioritized as an integral part of universal contracts to bring about a green and circular economy. Emerging supplementary encouraging regulation desires to line up fossil fuel-based resources instantly for the future. Indeed, bio-based polymeric plastics are evolving as a viable solution at an early stage of virgin plastic pollution. However, this biodegradable plastic market share is below 2% in 2017 [119]. The marginal market share of bio-based plastic typically could be due to cheap fossil-based plastics resources, the huge land use requirement and capital, and the immature reprocessing, and/or dumping methods [120]

Thus move to a circular economy involves change direction in the whole plastics value chain starting from resource and production to material design, reprocessing possibilities after their worthwhile lifespan as shown in Fig. 4. Biomass by-products appear to be a likely option to increase bio-based plastic in bio-refinery because it overwhelms land use requirements and enhances production [120]. For effective utilization of by-products, massive efforts must be exerted to develop superior hydrolytic ability microbial strains that would permit direct conversion of biomass and the extraction of value-added products intended for the synthesis process [120]. Soon, high-performance bio-based polymer associated with enhanced sustainable end-life alternatives arises to be the natural environment is sustainable.

The improvement of bio-based progressive with similar physical properties to fuel-based complements would be important for the PPE during pandemic scenarios. The bio-based polymer would let reuse or processing resulted in enhancing energy recovery in the prevailing waste-to-energy setup later appropriate decontamination on the natural environment.

On the other hand, reusing and recycling through mechanical and chemical upcycling are sustainable alternatives for discarded plastic handling. For example, the

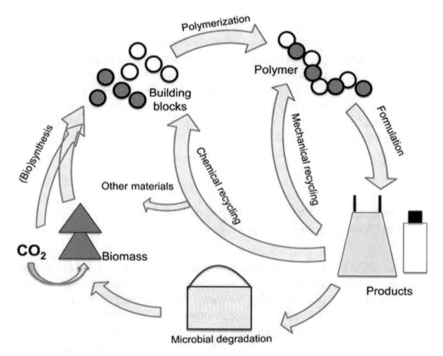

Fig. 4 Demonstration of the production of bio-based plastics and their recycling strategies [120] adapted with permission of Elsevier, 2020

plastic wastes can be mechanically shredded into pellets for recycle to the local market [104].

Since PPE and SUPs are polypropylene (PP), repurposing after proper disinfection can effectively reduce the waste volume to energy overcome the current plastic pollution scenario. Thermochemical degradation of these wastes converts to value-added products such as biofuels. For instance, Aragaw and Mekonnen proposed pyrolysis as an effective alternative to mitigation measure for the current pandemic plastic waste particularly faces mask and gloves, and achieved 75 wt% conversions into bio-crude oil [121]. Similarly, Jung et al. valorized of face mask to syngas and proposed a naturally non-threatening disposal procedure, concurrently realizing the production of such valuable fuels from the face mask [122]. Likewise, Chew et al. proposed hazardous wastes for potential pyrolysis and co-pyrolysis feedstock for future green environment and circular economy [123]. Hence, the growth of bio-based and decomposable solutions for SUPs and biomedical wastes would reduce their environmental load through landfilling or waste-to-energy options.

4.4 Implementation for the Rational Use of PPE

Single-use PPE is not a viable practice, and multidisciplinary practical skill is essential to deal with the PPE litter problem and proposing proper sustainable solutions. Numerous suggestions for the most effective use of PPE have been anticipated by WHO (Interim guidance, Feb 27, 2020). Among these sustainable implementations, the use of social distancing, the encouragement of telemedicine, avoiding overcrowdedness, telemarketing, and online/teleshopping minimizes the need for PPE thereby reduced wastes. In addition, it is also vital to select high-grade PPE with can be potentially sterilized and reused after proper disinfection. Much of the evidence has also shown that PPE sanitization and reuse are conceivable in all-encompassing through decontamination techniques in the face of PPE shortage [124]. This coherent use and reuse of PPE resources decrease in the generation of medical plastic wastes.

5   Conclusion and Future Directives

The impact of novel coronavirus (COVID-19) pandemic on the environment, economy and society are become a hot spot and attracted the researchers and governments. WHO declared that the COVID-19 pandemic will continue for an extended period, and all countries are occupied on prevention policies on health and environmental care. As a result, the extent of the pandemic spread worldwide will also have a prolonged effect on environmental sustainability. Thus, indefensible environmental management attributes to increased transmission of diseases from animal to human beings. The pandemic outbreak augments extensive use of SUPs and quantity of infectious COVID-19-related medical waste in victim nations. The end-users behavior has been shifted to SUPs and PPE preference for virus protection with less attention to the disposable plastic wastes to the environment. These plastic utilization custom modifications have without exemption lead to plastic waste proper management challenges and a setback of policies on source reduction of SUPs products. The confinement and lockdown policies perhaps the volume of food waste greater than before but perhaps to recognize significance of food waste reduction. However, the higher food waste in the households has been outweighed by a reduction in municipal solid wastes centers.

The pandemic has inspired extra care on the interaction of humans and the natural environment in addition to modification of lifestyles and consumption patterns. Furthermore, the pandemic changes human and environmental integration while the negative impact will surpass the opportunities in the future. Consequently, the huge production and use of SUPs and PPE imposed by the pandemic makes plastic pollution endless in the environment. To tackle this plastic pollution, the circular economy strategy should be progress for SUPs and PPE controlling during and later the current pandemic.

In conclusion, the COVID-19 pandemic has opportunities in improved the water and air quality index as well as ecological renovation. However, the pandemic restricted solid waste management and added extra plastic litter to the existing globally. Therefore, emphasizing the production, use and correct disposal is vital these days as PPEs will remain to be in high demand. In this manner, researchers with motivated efforts are needed for novel PPE and SUPs materials that would be reused and/or redesign for biodegradable plastic production and other reusable fabrics to reduce environmental pollution as well as implementing plastic policy guidance at the global level for environmental sustainability.

6 Conflict of Interest

The authors declare that they have no known competing financial interests or personal relationships that could have appeared to influence the work reported in this paper.

Acknowledgments The authors want to thank the previously published paper writers. Also, want to apologize to all intellectuals, and organizations whose involvement in the field may have been reviewed by the mistake or inadequately recognized.

Authors Contribution Statements

Bassazin Ayalew Mekonnen Conceptualization, Investigation, Formal analysis, Writing–original draft.

Tadele Assefa Aragaw Supervision, Validation, Formal analysis, Writing–review and editing

References

1. Arora NK, Mishra J (2020) COVID-19 and importance of environmental sustainability. Environ Sustain 3:117–119. https://doi.org/10.1007/s42398-020-00107-z
2. Zambrano-Monserrate MA, Alejandra M, Sanchez-alcalde L (2020) effects of COVID-19 on the environment. Sci Total Environ 728:138813. https://doi.org/10.1016/j.scitotenv.2020. 138813
3. Chakraborty I, Maity P (2020) COVID-19 outbreak: Migration, effects on society, global environment and prevention. Sci Total Environ 728:138882. https://doi.org/10.1016/j.scitot env.2020.138882
4. Rume T, Islam SMDU (2020) Environmental effects of COVID-19 pandemic and potential strategies of sustainability. Heliyon 6:e04965. https://doi.org/10.1016/j.heliyon.2020.e04965
5. Patrício Silva AL et al (2021) Increased plastic pollution due to COVID-19 pandemic: Challenges and recommendations. Chem Eng J 405:126683. https://doi.org/10.1016/j.cej.2020. 126683
6. Grodzińska-Jurczak M, Krawczyk A, Jurczak A, Strzelecka M, Rechciński M, Boćkowski M (2020) Environmental choices vs. COVID-19 pandemic fear – plastic governance re-assessment. Soc Regist 4:49–66. https://doi.org/10.14746/sr.2020.4.2.04.
7. Ammendolia J, Saturno J, Brooks AL, Jacobs S, Jambeck JR (2021) An emerging source of plastic pollution: Environmental presence of plastic personal protective equipment (PPE) debris related to COVID-19 in a metropolitan city. Environ Pollut 269:116160. https://doi. org/10.1016/j.envpol.2020.116160
8. Fadare OO, Okoffo ED (2020) Covid-19 face masks: A potential source of microplastic fibers in the environment. Sci Total Environ 737:140279. https://doi.org/10.1016/j.scitotenv.2020. 140279
9. Aragaw TA (2020) Surgical face masks as a potential source for microplastic pollution in the COVID-19 scenario. Mar Pollut Bull 159:111517. https://doi.org/10.1016/j.marpolbul.2020. 111517
10. Cheval S, Adamescu CM, Georgiadis T, Herrnegger M, Piticar A, Legates DR (2020) Observed and Potential Impacts of the COVID-19 Pandemic on the Environment. Int J Environ Res Public Health 17:4140
11. Eroğlu H (2020) Efects of Covid-19 outbreak on environment and renewable energy sector. Environ Dev Sustain 28:1–9
12. Espejo W, Celis JE, Chiang G, Bahamonde P (2020) Environment and COVID-19: Pollutants, impacts, dissemination, management and recommendations for facing future epidemic threats. Sci Total Environ 747:141314. https://doi.org/10.1016/j.scitotenv.2020.141314
13. Saadat S, Rawtani D, Hussain CM (2020) Environmental perspective of COVID-19. Sci Total Environ 728:138870. https://doi.org/10.1016/j.scitotenv.2020.138870
14. Tobías A (2020) Evaluation of the lockdowns for the SARS-CoV-2 epidemic in Italy and Spain after one month follow up. Sci Total Environ 725:138539. https://doi.org/10.1016/j.sci totenv.2020.138539
15. Wang Q, Su M (2020) A preliminary assessment of the impact of COVID-19 on environment: a case study of China. Sci Total Environ 728:138915. https://doi.org/10.1016/j.scitotenv.2020. 138915
16. Dantas G, Siciliano B, Boscaro B, Cleyton M, Arbilla G (2020) The impact of COVID-19 partial lockdown on the air quality of the city of Rio de Janeiro, Brazil. Sci Total Environ 729:139085. https://doi.org/10.1016/j.scitotenv.2020.139085
17. Liang Y, Song Q, Wu N, Li J, Zhong Y, Zeng W (2021) Repercussions of COVID-19 pandemic on solid waste generation and management strategies. Front Environ Sci Eng 15:1–18. https:// doi.org/10.1007/s11783-021-1407-5
18. WHO (2016) https://www.who.int/health-topics/air-pollution#tab=tab_1. Accessed date 14 March 2021
19. Zhang Q et al (2017) Transboundary health impacts of transported global air pollution and international trade. Nature 543:705–709. https://doi.org/10.1038/nature21712

20. Forster PM et al (2020) Current and future global climate impacts resulting from COVID-19. Nat Clim Chang 10:913–919. https://doi.org/10.1038/s41558-020-0883-0
21. Tobías A et al (2020) Changes in air quality during the lockdown in Barcelona (Spain) one month into the SARS-CoV-2 epidemic. Sci Total Environ 726:138540. https://doi.org/10. 1016/j.scitotenv.2020.138540
22. Caine P (2020) Environmental impact of COVID-19 lockdowns seen from space. Sci Nat 2.
23. Wang P, Chen K, Zhu S, Wang P, Zhang H (2020) Severe air pollution events not avoided by reduced anthropogenic activities during COVID-19 outbreak. Resour Conserv Recycl 158:104814. https://doi.org/10.1016/j.resconrec.2020.104814
24. Muhammad S, Long X, Salman M (2020) COVID-19 pandemic and environmental pollution: A blessing in disguise? Sci Total Environ 728:138820. https://doi.org/10.1016/j.scitotenv. 2020.138820
25. EEA (2020) Air pollution goes down as Europe takes hard measures to combat Coronavirus. European Environmental Agency (EEA), Copenhagen. https://www.eea.europa.eu/highlights/ air-pollution-goes-down-as. Accessed date 14 April 2020
26. Berman JD, Ebisu K (2020) Changes in U.S. air pollution during the COVID-19 pandemic. Sci Total Environ 739:139864. https://doi.org/10.1016/j.scitotenv.2020.139864
27. Adams MD (2020) Air pollution in Ontario, Canada during the COVID-19 State of Emergency. Sci Total Environ 742:140516. https://doi.org/10.1016/j.scitotenv.2020.140516
28. Nakada LYK, Urban RC (2020) COVID-19 pandemic: Impacts on the air quality during the partial lockdown in São Paulo state, Brazil. Sci Total Environ 730:139087. https://doi.org/10. 1016/j.scitotenv.2020.139087
29. Kerimray A et al (2020) Assessing air quality changes in large cities during COVID-19 lockdowns: the impacts of traf fi c-free urban conditions in Almaty, Kazakhstan. Sci Total Environ 730:139179. https://doi.org/10.1016/j.scitotenv.2020.139179
30. Baldasano JM (2020) COVID-19 lockdown effects on air quality by NO 2 in the cities of Barcelona and Madrid (Spain). Sci Total Environ 741:140353. https://doi.org/10.1016/j.sci totenv.2020.140353
31. Otmani A et al (2020) Impact of Covid-19 lockdown on PM 10, SO 2 and NO 2 concentrations in Salé City (Morocco) in Salé city. Sci Total Environ 735:139541. https://doi.org/10.1016/ j.scitotenv.2020.139541
32. Thiessen T (2020) How clean air cities could outlast COVID-19 lockdowns.
33. India Environment Portal (IEP) (2020) Impact of lockdown (25th March to 15th April) on air quality. http://www.indiaenvironmentportal.org.in/content/467415/impact-oflockdown-25th-march-to-15th- april-on-air-quality/. Accessed 4 August 2020.
34. Henriques M (2020) Will Covid-19 have a lasting impact on the environment? BBC News, London
35. Zogopoulos E (2020) COVID-19: the curious case of a green virus. Energy industry review, 17 April 2020
36. Wallace G (2020) Airlines and TSA report 96% drop in air travel as pandemic continues. CNN, 9 April 2020
37. IEA (2020) Oil market report: March 2020. The International Energy Agency, Paris
38. CREA (2020) Air quality improvements due to COVID-19 lock-down in India. Centre for Research on Energy and Clean Air
39. Ghosh I (2020) The emissions impact of coronavirus lockdowns, as shown by satellites
40. Myllyvirta L (2020) Coronavirus temporarily reduced China's CO_2 emissions by a quarter. Carbon Brief. https://www.carbonbrief.org/analysis-coronavirus-has-temporarily-reduced-chinas-co2-emissions-by-a-quarter
41. Evans S (2020) Global emissions analysis: coronavirus set to cause largest ever annual fall in CO_2 emissions. Carbon brief, 4 September 2020
42. Le Quéré C et al (2020) Temporary reduction in daily global CO 2 emissions during the COVID-19 forced confinement. Nat Clim Chang 10:647–653. https://doi.org/10.1038/s41 558-020-0797-x

43. Dutheil F, Baker JS, Navel V (2020) COVID-19 as a factor influencing air pollution? Environ Pollut 263:2019–2021. https://doi.org/10.1016/j.envpol.2020.114466

44. Faridi S et al (2020) A field indoor air measurement of SARS-CoV-2 in the patient rooms of the largest hospital in Iran. Sci Total Environ 725:1–5. https://doi.org/10.1016/j.scitotenv.2020.138401

45. The Environmental Protection Bureau (DSPA) Macao SAR, "Enviormental data.," 2020.

46. Ouhsine O, Ouigmane A, Aba B, Isaifan RJ, Berkani M (2020) Impact of COVID-19 on the qualitative and quantitative aspect of household solid waste. Glob. J. Environ. Sci. Manag. 6:41–52. https://doi.org/10.22034/GJESM.2019.06.SI.05

47. AMSA (2020) Waste management and cleaning services in Milan during COVID-19

48. Ragazzi M, Cristina E, Schiavon M (2020) Municipal solid waste management during the SARS-COV-2 outbreak and lockdown ease: Lessons from Italy. Sci Total Environ 745:141159. https://doi.org/10.1016/j.scitotenv.2020.141159

49. ACR+ (2020) Municipal waste management and COVID-19. Association of Cities and Regions for Sustainable Resource Management

50. Somani M, Srivastava AN, Gummadivalli SK, Sharma A (2020) Indirect implications of COVID-19 towards sustainable environment: An investigation in Indian context. Bioresour. Technol. Reports 11:100491. https://doi.org/10.1016/j.biteb.2020.100491

51. Tsukiji M, et al (2020) Waste management during the COVID-19 pandemic

52. ADEPT (2020) COVID 19: waste survey results. Week Commencing, 18 May

53. Zand AD, Heir AV (2020) Emerging challenges in urban waste management in Tehran, Iran during the COVID-19 pandemic. Resour Conserv Recycl 162(June):105051. https://doi.org/10.1016/j.resconrec.2020.105051

54. Yunus AP, Masago Y, Hijioka Y (2020) COVID-19 and surface water quality: Improved lake water quality during the lockdown. Sci Total Environ 731:139012. https://doi.org/10.1016/j.scitotenv.2020.139012

55. Singhal S, Matto M (2020) COVID-19 lockdown: a ventilator for rivers. DownToEarth. In: Somani M, et al (eds), Bioresource Technology Reports, 11, p. 100491. https://www.downtoearth.org.in/blog/covid-19-lockdown-aventilator-for-rivers-70771. (Accessed 20 May 2020)

56. UPCB (2020) Uttarakhand Pollution Control Board (UPSC), 2020. Water Quality during Lockdown Period. Government of Uttarakhand, India

57. BIS (2020) Bureau of Indian Standards Drinking Water Specifications. BIS 10500:2012. BIS, New Delhi, India

58. Arif M, Kumar R, Parveen S (2020) Reduction in water pollution in Yamuna River due to lockdown under COVID-19 Pandemic. ChemRxiv 9(12):84–89. https://doi.org/10.26434/chemrxiv.12440525.v1

59. Cripps K (2020) Thailand's most popular island goes into lockdown as Covid-19 cases surge. CNN travel, CNN, 10 April 2020

60. Clifford C (2020) The water in Venice, Italy's canals is running clear amid the COVID-19 lockdown

61. Kundu C (2020) Has the Covid-19 lockdown returned dolphins and swans to Italian waterways? The India Today, 22 March 2020

62. Rahman M, Griffiths MD, Mamun MA (2020) Correspondence Biomedical waste amid COVID-19: Perspectives from Bangladesh. Lancet Glob Heal 20:30349. https://doi.org/10.1016/S2214-109X(20)30349-1

63. Jribi S, Ben H, Doggui D, Debbabi H (2020) COVID-19 virus outbreak lockdown: What impacts on household food wastage ? Environ Dev Sustain 22:3939–3955. https://doi.org/10.1007/s10668-020-00740-y

64. Cooper R (2020) Water security beyond Covid-19

65. Goines L, Hagler L (2007) Noise pollution: a modem plague. South Med J 100:287–294

66. Sims J (2020) Will the world be quieter after the pandemic? BBC Future

67. WHO (2012) Global estimates on prevalence of hearing loss mortality and burden of diseases and prevention of blindness and deafness

68. Solan M, Hauton C, Godbold JA, Wood CL, Leighton TG, White P (2016) Anthropogenic sources of underwater sound can modify how sediment-dwelling invertebrates mediate ecosystem properties. Sci Rep 6:20540. https://doi.org/10.1038/srep20540

69. Gandhiok J, Ibra M (2020) Covid-19: noise pollution falls as lockdown rings in sound of silence. The Times of India, 23 April 2020

70. SIMS (2020) Will the world be quieter after the pandemic?

71. Lenzen M, Sun Y, Faturay F, Ting Y, Geschke A, Malik A (2018) The carbon footprint of global tourism. Nat Clim Chang 8:522–528. https://doi.org/10.1038/s41558-018-0141-x

72. Puig R, Kiliç E, Navarro A, Albertí J, Chacón L, Fullana-i-palmer P (2017) Environment Inventory analysis and carbon footprint of coastland-hotel services: a Spanish case study. Sci Total Environ 595:244–254. https://doi.org/10.1016/j.scitotenv.2017.03.245

73. Pereira RPT, Ribeiro GM, Filimonau V (2017) The carbon footprint appraisal of local visitor travel in Brazil: A case of the Rio de Janeiro-São Paulo itinerary. J Clean Prod 141:256–266. https://doi.org/10.1016/j.jclepro.2016.09.049

74. Islam SMD, Bhuiyan MA (2018) Sundarbans mangrove forest of Bangladesh: causes of degradation and sustainable management options. Environ Sustain 1:113–131. https://doi.org/10.1007/s42398-018-0018-y

75. Corlett RT et al (2020) Impacts of the coronavirus pandemic on biodiversity conservation. Biol Conserv 246:108571. https://doi.org/10.1016/j.biocon.2020.108571

76. Rahman M (2020) Rare dolphin sighting as Cox' s Bazar lockdown under COVID-19 coronavirus

77. Abu-Qdais HA, Al-Ghazo MA, Al-Ghazo EM (2020) Statistical analysis and characteristics of hospital medical waste under novel Coronavirus outbreak. Glob. J. Environ. Sci. Manag. 6:21–30. https://doi.org/10.22034/GJESM.2019.06.SI.03

78. Klemeš JJ, Van Fan Y, Tan RR, Jiang P (2020) Minimising the present and future plastic waste, energy and environmental footprints related to COVID-19. Renew Sustain Energy Rev 127(April):109883. https://doi.org/10.1016/j.rser.2020.109883

79. Ministry of Environment of South Korea (MoE Korea) (2020). Press Release, Minister of Environment, Check the Management of Wastes Related to COVID-19. me.go.kr/home/web/board/read.do;jsessionid=RwDT9rf77AYJRhB3t87es+

80. Asian Development Bank (ADB) (2020) Managing infectious medical waste during the COVID-19 pandemic.

81. Singh N, Tang Y, Ogunseitan OA (2020) Environmentally Sustainable Management of Used Personal Protective Equipment. Environ Sci Technol 54:8500–8502. https://doi.org/10.1021/acs.est.0c03022

82. Prata JC, Silva AL, Walker TR, Duarte AC, Rocha-Santos T (2020) COVID-19 Pandemic Repercussions on the Use and Management of Plastics. Environ Sci Technol 54(13):7760–7765. https://doi.org/10.1021/acs.est.0c02178

83. Adyel TM (2020) Accumulation of plastic waste during COVID-19". Science 369:1314–1315

84. Bengali S (2020) The COVID-19 pandemic is unleashing a tidal wave of plastic waste. The Los Angeles Times

85. Calma J (2020) The COVID-19 pandemic is generating tons of medical waste. The Verge, 26 March 2020

86. Zuo M (2020) Coronavirus leaves China with mountains of medical waste,' South China Morning Post. https://www.scmp.com/news/china/society/article/3074722/coronavirus-leaves-china-mountains-medical-waste

87. Allison A, et al (2020) The environmental dangers of employing single-use face masks as part of a COVID-19 exit strategy

88. Xanthos D, Walker TR (2017) International policies to reduce plastic marine pollution from single-use plastics (plastic bags and microbeads): A review. Mar Pollut Bull 118:17–26. https://doi.org/10.1016/j.marpolbul.2017.02.048

89. De-la-Torre GE, Rakib MRJ, Pizarro-Ortega CI, Dioses-Salinas DC (2021) Occurrence of personal protective equipment (PPE) associated with the COVID-19 pandemic along the coast of Lima, Peru. Sci Total Environ 774:145774. https://doi.org/10.1016/j.scitotenv.2021.145774

90. Jambeck JR et al (2015) Plastic waste inputs from land into the ocean". Science 347:768–771
91. Lebreton LCM, Van Der Zwet J, Damsteeg JW, Slat B, Andrady A, Reisser J (2017) River plastic emissions to the world's oceans. Nat Commun 8:15611. https://doi.org/10.1038/ncomms15611
92. Eriksen M et al (2013) Microplastic pollution in the surface waters of the Laurentian Great Lakes. Mar Pollut Bull 77:177–182. https://doi.org/10.1016/j.marpolbul.2013.10.007
93. Monteiro RCP, Ivar do Sul JA, Costa MF (2018) Plastic pollution in islands of the Atlantic Ocean. Environ Pollut 238:103–110. https://doi.org/10.1016/j.envpol.2018.01.096
94. Wright SL, Thompson RC, Galloway TS (2013) The physical impacts of microplastics on marine organisms: a review. Environ Pollut 178:483–492. https://doi.org/10.1016/j.envpol.2013.02.031
95. Connors EJ (2017) Distribution and biological implications of plastic pollution on the fringing reef of Mo'orea, French Polynesia. PeerJ 5:e3733. https://doi.org/10.7717/peerj.3733
96. Aragaw TA, Mekonnen, BA (2021) Distribution and impact of microplastics in the aquatic systems: a review of ecotoxicological effects on biota. In: Muthu SS (ed) Microplastic pollution. sustainable textile: production, processing, manufacturing and chemistry. Springer, Singapore, pp 65–104.
97. Ma B, Li X, Jiang Z, Jiang J (2018) Recycle more, waste more? When recycling efforts increase resource consumption. J Clean Prod 206:870–877. https://doi.org/10.1016/j.jclepro.2018.09.063
98. Vanapalli KR et al (2021) Challenges and strategies for effective plastic waste management during and post COVID-19 pandemic. Sci Total Environ 750:141514. https://doi.org/10.1016/j.scitotenv.2020.141514
99. De-la-Torre GE, Aragaw TA (2021) What we need to know about PPE associated with the COVID-19 pandemic in the marine environment. Mar Pollut Bull 163:111879. https://doi.org/10.1016/j.marpolbul.2020.111879
100. Ahmed W et al (2020) First confirmed detection of SARS-CoV-2 in untreated wastewater in Australia: A proof of concept for the wastewater surveillance of COVID-19 in the community. Sci Total Environ 728:138764. https://doi.org/10.1016/j.scitotenv.2020.138764
101. Nghiem LD, Morgan B, Donner E, Short MD (2020) The COVID-19 pandemic: Considerations for the waste and wastewater services sector. Case Stud Chem Environ Eng 1:100006. https://doi.org/10.1016/j.cscee.2020.100006
102. Mallapaty S (2020) How sewage could reveal true scale of coronavirus outbreak. Nature 580:176–177
103. Boyle L (2020) Bird dies after getting tangled in coronavirus face mask. Independent New York
104. Williams-Wynn MD, Naidoo P (2020) A review of the treatment options for marine plastic waste in South Africa. Mar Pollut Bull 161:111785. https://doi.org/10.1016/j.marpolbul.2020.111785
105. Ferraro G, Failler P (2020) Governing plastic pollution in the oceans: Institutional challenges and areas for action. Environ Sci Policy 112:453–460. https://doi.org/10.1016/j.envsci.2020.06.015
106. Yang Y, Liu W, Zhang Z, Grossart HP, Gadd GM (2020) Microplastics provide new microbial niches in aquatic environments. Appl Microbiol Biotechnol 104(15):6501–6511. https://doi.org/10.1007/s00253-020-10704-x
107. Di Martino M et al (2020) Elective surgery during the SARS-CoV-2 pandemic (COVID-19): a morbimortality analysis and recommendations on patient prioritisation and security measures. Cir Esp 98(9):525–532. https://doi.org/10.1016/j.ciresp.2020.04.029
108. NREB20-02 (2020) Corona virus waste: disposal of face masks. Environment Bulletin
109. Robaina M, Murillo K, Rocha E, Villar J (2020) Circular economy in plastic waste: efficiency analysis of European countries. Sci Total Environ 730:139038. https://doi.org/10.1016/j.scitotenv.2020.139038
110. Selvaranjan K, Navaratnam S, Rajeev P, Ravintherakumaran N (2021) Environmental challenges induced by extensive use of face masks during COVID-19: a review and potential solutions. Environ. Challenges 3:100039. https://doi.org/10.1016/j.envc.2021.100039

111. Layt S (2020) Queensland researchers hit sweet spot with new mask material. The Age
112. Ho S (2020) Vietnamese company creates world's first biodegradable coffee face mask. greenqueen
113. Staff R (2020) From field to compost: French firm develops hemp face masks. Reuters
114. Lackner M (2015) Bioplastics: biobased plastics as renewable and/or biodegradable alternatives to petroplastics. In: Othmer K (ed) Kirk-Othmer encyclopedia of chemical technology, 6th ed. Wiley, New York, p 3
115. Patrício Silva PL et al (2020) Rethinking and optimising plastic waste management under COVID-19 pandemic: Policy solutions based on redesign and reduction of single-use plastics and personal protective equipment". Sci Total Environ 742:140565. https://doi.org/10.1016/j.scitotenv.2020.140565
116. Chin AWH et al (2020) Stability of SARS-CoV-2 in different environmental conditions. The Lancet Microbe 1:e10. https://doi.org/10.1016/s2666-5247(20)30003-3
117. Ilyas S, Srivastava RR, Kim H (2020) Disinfection technology and strategies for COVID-19 hospital and bio-medical waste management. Sci Total Environ 749:141652. https://doi.org/10.1016/j.scitotenv.2020.141652
118. ACR+ (2020) Association of cities and regions for sustainable resource management
119. PlasticsEurope (2019) Plastics: the facts 2019. An analysis of European plastics production, demand and waste data. PlasticsEurope Brussels, Belgium
120. Hatti-Kaul R, Nilsson LJ, Zhang B, Rehnberg N, Lundmark S (2020) Designing Biobased Recyclable Polymers for Plastics. Trends Biotechnol 38:50–67. https://doi.org/10.1016/j.tibtech.2019.04.011
121. Aragaw TA, Mekonnen BA (2021) Current plastics pollution threats due to COVID-19 and its possible mitigation techniques: a waste-to-energy conversion via Pyrolysis. Environ Syst Res 10:8. https://doi.org/10.1186/s40068-020-00217-x
122. Jung S, Lee S, Dou X, Kwon EE (2021) Valorization of disposable COVID-19 mask through the thermo-chemical process. Chem Eng J 405:126658. https://doi.org/10.1016/j.cej.2020.126658
123. Chew KW, Chia SR, Chia WY, Cheah WY, Munawaroh HSH, Ong WJ (2021) Abatement of hazardous materials and biomass waste via pyrolysis and co-pyrolysis for environmental sustainability and circular economy. Environ Pollut 278:16836
124. Rubio-Romero JC, Pardo-Ferreira MC, Torrecilla-García JA, Calero-Castro S (2020) Disposable masks: Disinfection and sterilization for reuse, and non-certified manufacturing, in the face of shortages during the COVID-19 pandemic". Saf Sci 129:104830. https://doi.org/10.1016/j.ssci.2020.104830

Healthy Sustainable Cities and the COVID-19 Pandemic: A Sustainable Development Goals Perspective

Stephane Louise Boca Santa, Graziela Oste Graziano Cremonezi, Thiago Coelho Soares, André Borchardt Deggau, and José Baltazar Salgueirinho Osório de Andrade Guerra

Abstract The world is experiencing a pandemic caused by the SARS-CoV-2 virus, which develops the disease COVID-19. Taking care of yourself and others are attitudes considered more than a norm or suggestion, it is an attitude governed by common sense and also, an act of love. A new reality for many: work at home or with protective equipment; home studies; limitations when leaving home (use of mask, alcohol gel, and temperature measurement in different environments); excessive hygiene; numerous surveys (products, services, medicines, vaccines); hospital products and equipment. The importance and benefits of living in a sustainable city have become even more evident. Investments and emergency aid, water management, waste management, home education, health, are old needs, even more evident in a new reality. A sustainable city aims to develop responsibly, taking into account the triple bottom (economic, social, environmental). Based on this context, the objective of this research is: How the concept of sustainable cities can contribute in a pandemic context. This research aims to contribute theoretically, by addressing a series of measures and actions, foreseen in a sustainable city and that can positively help the population, when they find themselves in a pandemic. Still, there is a practical and social contribution, as the research contributes to the management of cities, aiming for smart, healthy and sustainable cities, focusing on education, security, and public health.

S. L. B. Santa · G. O. G. Cremonezi · T. C. Soares · A. B. Deggau
University of Southern Santa Catarina (UNISUL), Florianópolis 88010-010, Brazil
e-mail: graziela.alano@unisul.br

T. C. Soares
e-mail: thiago.soares@unisul.br

J. B. S. O. de Andrade Guerra (✉)
Centre for Sustainable Development (GREENS), University of Southern Santa Catarina,
Florianópolis 88010-010, Brazil
e-mail: jose.baltazarguerra@animaeducacao.com.br

Cambridge Centre for Environment, Energy and Natural Resource Governance (CEENRG),
University of Cambridge, Cambridge CB2 1TN, UK

S. S. Muthu (ed.), *COVID-19*, Environmental Footprints and Eco-design
of Products and Processes, https://doi.org/10.1007/978-981-16-3860-2_6

Keywords Healthy cities · Sustainable cities · Smart cities · COVID-19 ·
Pandemic impact · Sustainable development goals · Triple bottom · Management
of cities

1 Introduction

There are several transformations and changes underway in the world, of varied
nature: socioeconomic, environmental, technological, and cultural change are accompanied by intense urbanization and a strong feeling that there are crises in governance
and values, as well as a lack of a forward-looking vision that is capable to meet the
new challenges that appear in such a fast environment (GPS 2013).

With an analysis of the information found in the "Guia da Gestão Pública Sustentável" (GPS), which stands for a Guide for Sustainable Public Management, Brazil
fits that picture perfectly: one of the most challenging questions this country faces
today is how to balance economic development, environmental sustainability and
social justice through transparent, and democratic governance.

The GPS Guide addresses this question systemically and horizontally. There is
no way to solve these problems separately, or to abandon one issue momentarily
and come back to it later: the socioeconomic crisis needs to be solved within the
natural limits imposed by the planet, reverting tendencies such as extreme climate
change and the depletion of natural resources, while not overlooking the importance
of solving social inequalities. Such solutions need to take into account that there is
urgency in the temporal dimension, and there is no time to lose.

The GPS Guide is based on the sustainable development goals (SDGs) created
by the United Nations (UN) in 2015 with the approval of its 192 countries to guide
policy-makers into creating a more sustainable planet for future generations. The
17 goals set up by this Agenda are composed of 169 indicators and cover social,
environmental, and economic development.

One of these goals is SDG 11: "make cities and human settlements inclusive,
safe, resilient, and sustainable" [52]. This goal cannot be reached on its own, since
several factors need to be considered, and issues such as energy, water, health, and
environmental protection connect SDG 11 to the other SDGs.

To use transparent and participatory processes to achieve SDG 11 in Brazil, the
GPS [23] has presented a sustainable cities program. This program is to be constantly
under construction, but its starting point is the assumption that a new paradigm of
development is possible, necessary, and urgent. The sustainable cities program seeks
to attach the idea of change to ongoing experiences and opportunities that can be
useful in the building of a new development model.

Sustainability is a broad concept, and its definition demands holistic, and multidisciplinary views. In order to measure how sustainable or unsustainable something
is, a series of indicators, criteria, and goals are generally used. This logic also applies
to sustainable cities: such a city needs to have sustainability as the premise of its
many actions, operations, services, and infrastructure.

In the effort to reach sustainable cities, specific terminology has emerged, bringing innovative concepts: a "compact city", emphasizing density; "eco-city", highlighting cultural and ecological diversity; "neotraditional development", seeking sustainable transport and land use; "urban contention", striving for containing urban expansion; "smart city", stressing cities that use technological innovation to create efficiency, sustainability, and quality of life; "healthy city", a concept crafted by the World Health Organization (WHO) underscoring human health; and "sustainable city", which provides a global view of operations and services in a city [8]. In this research, we will use the terms smart cities, sustainable cities, and healthy cities.

The GPS [23] recommends that modern urban planning must have a sustainability-based framework that encompasses areas such as economic, cultural, social, ecological, technological, demographic, and tributary. Such a framework could provide for an integrated view of issues and connect local policy-making.

The strategic planning proposed by GPS [23] focuses on a systemic and participatory view, considering short, medium, and long-term projects—so that continuity will be assured, especially in infrastructure projects, which normally take longer. The planning also suggests that observable target are set, so that the public can transparently follow progress.

All of these problems already existed in early 2020, when the SARS-CoV-2 was identified in China. The COVID-19 pandemic that followed has had profound impacts on cities, by bringing the economy and social interactions to a halt, threatening livelihoods and jobs, made transport more difficult and crowded hospitals [38].

The COVID-19 pandemic spread swiftly around the world, creating broad and multidimensional repercussions on national and international dynamics, and posing a series of implications for human relations [42]. People and firms were forced to innovate and look for new ways of going about their activities, especially with more people working from home, which demanded adaptations in home infrastructures and improvements in hygiene.

Cities were promptly exposed and saw themselves in danger. For a long time, cities have been looking for resilient and sustainable solutions to a series of problems, ranging from fighting for urban spaces or political, financial, geographical, or environmental calamities, all, while managing their visible asymmetries and disparities [38]. The author adds that the fast-paced rhythm of change felt in the cities needs to be reversed in order to find new solutions and languages to create changes in cities.

Many of the adaptations that need to be made, presented in the concepts of healthy, smart, and sustainable cities, are a consensus among urban dwellers: the creation of green areas for sports and leisure, investments in hospital bed and support for health professionals, lower emission of pollutants, clean energy, access to quality housing, development of new technologies of information and better governance.

This chapter has as its proposal to draw a parallel between the needs imposed by the COVID-19 pandemic and the characteristics of sustainable cities to determine how a sustainable city could be effective in the context of a pandemic.

2   Literature Review

The literature review is divided into three themes: (2.1) The COVID-19 pandemic; (2.2) Healthy smart sustainable cities and their characteristics; and (2.3) the COVID-19 pandemic and the healthy smart sustainable cities.

2.1   The COVID-19 Pandemic

The Brazilian Ministry of Health [10] highlights that the coronaviruses are a known family of viruses observed in several animals, such as cattle, camels, cats, and bats. On rare occasions, viruses of this family have infected humans, but the MERS-Cov and the SARS-CoV were two examples. In December 2019, a new virus appeared: the SARS-CoV-2 was identified in China and provoked a disease that spread quickly, the COVID-19.

People infected with COVID-19 have presented different clinical pictures, ranging from asymptomatic infections to serious conditions. The World Health Organization (WHO) affirms that most of the patients (around 80%) will face few or no symptoms, and the other 20% of the patients will need to be admitted to the hospital facing symptoms such as breathing difficulty. The Brazilian Health Ministry [10] adds that the most common symptoms are similar to those observed in a common cold: coughing, fever, sore throat, breathing difficulty, temporary losses in the senses of taste and smell, and tiredness.

The SARS-CoV-2 is transmitted through respiratory droplets and is capable to lodge in several surfaces such as tables, chairs, walls, floors, and even medical equipment. In facing this pandemic, such surfaces must be cleaned regularly, especially in environments where the presence of the virus is known, such as hospitals [59]. When there is large-scale community transmission, there is a need to monitor hygiene in every environment.

The WHO [59] states that "contact transmission occurs when contaminated hands touch the mucosa of the mouth, nose, or eyes; the virus can also be transferred from one surface to another by contaminated hands, which facilitates indirect contact transmission". The transmission of the virus was associated with physical contact between individuals in closed environments—such as houses, health clinics, workplaces, schools, and others—so the recommendations for stopping the spread are to practice social distancing, using masks, washing hands frequently, and using alcohol gel.

The WHO [59] (p 1) further recommended to its member-states improvements in hygiene and "providing universal access to public hand hygiene stations and making their use obligatory on entering and leaving any public or private commercial building and any public transport facility", while "improving access to hand hygiene facilities and practices in health care facilities". The problem is that many cities, especially in low-income countries, find trouble in making hygiene products available for their citizens.

Such protocols and technical recommendations were widely shared among governments and populations alike, and different sectors of society have organized themselves to produce further information about how people can protect themselves. The sustainable cities program has suggested practices such as urban strategies; social assistance; employment and wages; public health; transparency and communication; and webs of support.

2.2 Healthy Smart Sustainable Cities and Its Characteristics

Cities play a crucial role in the global sustainable development agenda: more than 54% of the world's population live in cities, and it is expected that at least 60% of the world will be urban dwellers by 2030. The idea of urban sustainability is widely discussed in the literature and among international institutions, in order to help cities to overcome social, economic, and environmental challenges and provide quality of life for their citizens, while considering the concerns of future generations [7].

Fabris et al [20] understand cities as living organisms: they are in constant transformation under the influence of geographic, environmental, cultural, social, demographic, political, institutional, and economic shifts. It is hard to compare urban spaces in space and time since each city has specific challenges in their urbanization.

The term "sustainable city" has emerged in the 1990s, right after the first conceptualizations of "sustainability". At the time, environmentalists, economists, and activists around the world criticized the process of economic development and the living standards in developed countries, considering that those activities led to high levels of consumption and an exaggerated waste of natural resources, creating social imbalances and pollution on water and air [7].

Sustainable cities, as defined by Fabris et al [20], are the cities that prioritize the implementation of a set of practices and infrastructure that allows them to meet the goals set by the Brundtland Report—a reference on the meaning of sustainability—and the 2030 Agenda, which set the SDGs. These cities emphasize social and individual rights and are referred to, in literature, by several names: green cities, digital cities, smart cities, information cities, knowledge cities, resilient cities, low-carbon cities, eco-cities, and omnipresent cities.

Ferreira et al [20] classify sustainable cities as those which preserve their green areas, maintaining their original ecosystems in an urban setting. The authors highlight that greener cities provide their citizens with a better quality of life, since they preserve the quality of air and climate, also facilitating the recovery of water systems.

In this scenario, quality of life and socioenvironmental balance are extremely important for sustainable cities to develop [19]. For that reason, a culture that prioritizes the environmental questions and the preservation of species must be developed.

Reference [9] claims that there is no such thing as a sustainable city, but cities that seek for sustainability. Thus, sustainable cities are a progressive process of implementation of sustainability criteria, and this process demands a series of values, attitudes, and principles—in public and private spheres.

Sustainable cities are also connected with smart cities: [3] argue that a smart city needs to be efficient and sustainable, combining a high quality of life with an optimization of resources. The authors further claim that sustainability must be one of the focuses of smart cities.

Reference [17] claims that the concept of smart cities must consider concepts such as sustainability and quality of life—but also taking into account technological and informational components.

Smart cities are those that show clean transportation, the use of alternative energies, and intense use of technology [45]. Smart cities use information and communications technology (ICT) and other technologies to make their actions more effective and improving the quality of services (QoS) [8, 44, 62].

Reference [44] defends that the smartness or intelligence of a city is a desire to perfect it based on the sustainability triple-bottom-line of social, environmental, and economic realities. The authors believe that smart cities are based upon four pillars: institutional infrastructure, physical infrastructure, social infrastructure, and economic infrastructure.

The third world cities, as [9] calls them, tend to pollute, urbanize and destroy systematically their own fundamental environmental support. Such attitudes also worsen existing housing crises and slums sometimes appear in ecological sanctuaries and vital river basins.

The search for a sustainable city entails challenges to make the local economy dynamic, creative and sustainable, by finding alternatives to traditional economic growth that will reflect on the quality of life and welfare for the population [19].

Bento et al. [20] add that sustainable cities demand the conservation and maintenance of natural resources, considering the importance of developing territorial planning that fits the need of each town. Cities, while they are not natural ecosystems, are linked in a systemic and interdependent process, and this system needs new urban governance to solve its conflicts.

To complement this discussion, Table 1 presents the 12 thematic axes in the sustainable cities program, each with its description.

The sustainable cities program has as its goal the promote sustainability and welfare of societies. To measure the environmental impacts of a city, [9] proposes using "ecological footprint", a term first used by Willian Rees. This method focuses on calculating the amount of productive land that is used, in hectares. The calculation shows that many cities and countries demand a much larger area than they actually use, while also considering the transfer of costs and impacts to other regions in the country.

Another way to measure the efficiency and efficacy of sustainable policies and practices is suggested by [57]: using environmental and economic indicators. Urban sustainable development must act on the premise that its principles and applications need to be constantly updated, as societies and contexts are also always changing.

Table 1 The axes of the sustainable cities program

Axes	Description
Governance	Strengthening decision-making processes by promoting participatory democracy instruments
Common natural resources	Taking full responsibility to protect, preserve, and warrant balanced access to common natural resources
Equity, social justice, and culture of peace	Promoting inclusive and solidary communities
Local management for sustainability	Implementing efficient management that covers planning, execution, and evaluation
Urban planning and design	Recognizing the strategic role played by urban planning and design in dealing with environmental, social, economic, cultural, and health problems, for the benefit of all
Culture for sustainability	Developing cultural policies that respect and value cultural diversity, pluralism, and natural, built and immaterial heritage—while promoting memory preservation and the transmission of natural, cultural, and artistic heritage
Education for sustainability and quality of life	Integrate, in formal and informal education alike, values and abilities for a sustainable and healthy life
Dynamic, creative, and sustainable local economy	Supporting and creating conditions for a dynamic and creative local economy, ensuring access to employment while considering environmental concerns
Responsible consumption and lifestyle options	Adopting and providing a responsible and efficient use of resources, while incentivizing sustainable patterns of consumption and production
Better mobility and less traffic	Promoting sustainable mobility, recognizing the existing interdependence between transports, health, environment, and cities
Local action for health	Promoting and protecting health and welfare for all citizens
From local to global	Assuming global responsibilities for peace, justice, equity, sustainable development, environmental protection, and biodiversity

Adapted from Fabris et al. [19]

The sustainable cities program (2021) presents good practices that when applied can produce concrete results and inspire other cities. As an example, the city of Fortaleza, in northeastern Brazil, was able to reduce the time that citizens spend in traffic and public transportation by analyzing a large amount of data regarding displacement of people and vehicles. The city later became known as a reference in urban mobility by the mobilization of the civil society and managers on behalf of data transparency, reaching third place in the Brazilian rankings of smart cities in 2019.

Boston, in the United States, is another example mentioned by the program. The city has launched a program called Imagine Boston 2030 to draw a pathway for urban growth, considering inequalities, and climate change. The planning began with a survey of two years, with 15.000 Bostonians responding about demographic and economic changes in that region. The plan encompasses actions in the areas of housing, health, safety, security, education, economy, energy, open spaces, environment, transportation, technology, art, and culture.

2.3   The COVID-19 Pandemic and the Healthy Smart Sustainable Cities

This chapter aims to understand how the needs imposed by the COVID-19 pandemic can be met by the concept of sustainable cities. Table 2 presents analyses and proposals from actors of private social investment, philanthropy, and other sectors of civil society, as proposed by the association of social investors in Brazil [22]. These proposals refer specifically to 2021, a year in which the pandemic is ongoing, and touch areas such as citizenship, economy, education, generation of jobs and incomes, public management, housing, infancy, urban mobility, climate change, SDGs, sanitation, and health.

The COVID-19 pandemic has caused countless changes in the life of billions of people, as exemplified in Table 2. Lockdown policies, the suspension of classes in schools, and incentives to working from home have interrupted or modified daily commutes. In this context, several impacts on urban mobility, whether social, economic, or environmental, were observed [16].

Couto et al. [16] have further observed that Brazil has not developed significant actions in urban mobility based on the lessons that were obtained during the pandemic period, thus harming several sectors, one of which is transportation. It is noteworthy that as in other places, active mobility is gaining space among the Brazilian population, which can be observed in the increase in the sale of bicycles. However, this way of moving about stumbles upon the lack of basic infrastructure to consolidate itself in the Brazilian scenario.

The sustainable cities portal (2021) prepared a list of good practices carried out in Brazil and in some other countries during the pandemic, as an example: (1) considering libraries as fundamental services and encouraging reading for adults and children; (2) Kerala, in India, focused on social assistance, access to basic services, financial support and health management, thus achieving a good recovery rate for COVID-19; (3) The city of São Paulo organized itself in order to maintain social distance and recommence medical activities; (4) in Niterói, Rio de Janeiro, a rapid response group was created for matters related to COVID-19, working in several areas in an integrated manner; (5) in Jordan, the government provided a platform

Table 2 Sustainable Cities and the COVID-19 Pandemic

Axes	Goals	Contributions
Citizenship	Recognizing and empowering base organizations	A solidarity network, built with the contribution of private actors, will spur attention to communities during the pandemic. Local organizations know the territories, have direct contact with populations, and can deliver help. It is also important to understand how these organizations work, to keep supporting them after the pandemic
Citizenship	Fostering and qualifying spaces for popular participation and social control	Deliberative democracy is a way for people to participate and make concrete decisions, instead of simply legitimizing processes that come from governments. And, for that technical knowledge is needed so that a public agenda is built, not tied to political conveniences because those are normally temporary
Education	Qualifying connectivity, boosting the use of technology	The pandemic caused many public school students to experience a learning delay, as they did not have adequate access to remote education due to lack of equipment or internet
Education	Supporting managers in providing the necessary conditions for resuming presential lessons	The reopening of schools will be paid for in the first days of January. The new representatives will have to promote the conditions for the resumption of classes in 2021, considering not only the logistical issues of the beginning of the school year, but also computing the effects of the pandemic on learning rates, dropout rates, and inequality between students

(continued)

Table 2 (continued)

Axes	Goals	Contributions
Generation of labor and income	Implementing entrepreneurship and income generation programs in the suburbs	Many people lost their jobs and small businesses were closed during the pandemic. For the resumption of economic growth, this theme will be of fundamental importance for the development of cities. Injecting capital into actions aimed at generating income in the suburbs can be an opportunity for private social investment and philanthropy to leave the care model that we were all forced to operate with at first
Generation of labor and income	Fostering innovative and sustainable local businesses to solve social and environmental challenges	Incubating and accelerating creative, supportive, and green ventures; low-carbon emissions; focusing on building local productive arrangements from the perspective of the sustainable city can be a pathway forward
Municipal Public Management	Support city halls to deal with fiscal impacts of the pandemic	The crisis surrounding COVID-19 generated an economic downturn with rising unemployment and decreasing revenue. In this scenario, cities will have to find solutions to this challenge, in order to stimulate the economic recovery and make the necessary fiscal adjustment
Municipal Public Management	Encouraging and supporting the design of integrated policies based on the demands of each territory	The world today shows very clearly that territory matters: each territory has its specificities. Therefore, the suburbs must be included in the design of these policies considering several factors such as health, education, environment, housing, and culture
Municipal Public Management	Encouraging collaboration between federated entities, and providing tools that can be immediately used, and that can directly affect the decision making of local managers	It is extremely necessary to start the debate on sustainability with instruments such as the Pluriannual Plan and Master Plans and Land Use Plans

(continued)

Table 2 (continued)

Axes	Goals	Contributions
Municipal Public Management	Demanding priority for master plans	The way of life of children in cities has been increasingly restricted to closed spaces, for a variety of reasons—ranging from the feeling of insecurity in the public space to the families' lack of time. Inequalities of class, race, gender, disabilities, etc. Inequalities shape the structures and organizations of Brazilian cities, generating territorial segregation and further exacerbating inequalities, including for children and their families. It is necessary to implement policies that evenly distribute access and security to green areas and public spaces in the city
Health	Promoting structured and long-term investments	Health issues should not remain an emergency issue in 2021, but some structural investment is still required
Sustainable Development	Addressing SDGS	Governments are failing to address the SDGs, and therefore, we are not seeing significant advances

(continued)

for remote education and teacher training; (6) the city of Tubarão invested in the local production of ventilators and masks with a 3D printer in partnership with technology companies and universities; (7) New York city launches educational material in 26 languages to reach immigrants, the material presents guidelines for health, hygiene, and social isolation; (8) the government of the city of Recife creates a mobile application to guide the population.

3 Methodology

The methodology of this research is divided into literature review procedures (3.1) and indicator elaboration procedures (3.2), which are qualitative stages of the research. During the analysis of the literature, it was possible to arrive at the selected indicators.

Table 2 (continued)

Axes	Goals	Contributions
Basic sanitation	Establishing greater engagement and partnerships with public authorities within the scope of the new legal framework for basic sanitation	Currently, 35 million Brazilians do not have access to water and almost 100 million do not have access to sewage collection (36 municipalities in the country's 100 largest cities have less than 60% of the population with sewage collection). The positive impacts of universal sanitation for the sustainable development of cities are notable, especially in the areas of health, education, income generation, and the environment—not to mention the promotion of dignity and citizenship of populations in a situation of social vulnerability, which will gain access to better housing and health conditions. The challenge is great and the engagement and partnerships between public authorities, companies in the sector and civil society are essential to move forward and guarantee more sustainable cities and better quality of life for the population
Climate change	Supporting the creation of solutions aimed at populations subject to the impacts of climate change in a more intense way in cities	Environmental racism leads millions of people in large cities to suffer the consequences of climatic events, such as floods and landslides. Black populations are more vulnerable and more susceptible to these issues, since the way of production of cities is structurally racist, segregating, and excluding. If cities do not face this challenge, life in big cities will be increasingly unsustainable in environmental terms, threatening to kill millions of black and poor lives
Urban mobility	Support reorganization of the territorial occupation of cities that encourage short circuits	If people do not have to travel daily for long distances to go to major centers to use public services, work, study, buy or visit a doctor, the overcrowding of public transport will be significantly reduced

Source GIFE [22]

3.1 Literature Review Procedures

A systematic review was chosen as the method to approach the literature. For the theoretical framework, articles collected in international databases are used. In the selection of the portfolio, the keywords for the research are defined as: sustainable cities; green cities; smart cities, healthy cities, and sustainable development in cities. The search used the Boolean expression: (sustainab * AND city) OR (green AND city) OR (city AND environment*) OR (smart AND city) OR (healthy AND city).

Then, the raw database was formed, so the next step is filtering the data. The first filter refers to redundancy: in this filter, all duplicate articles are excluded. The second filter refers to the title alignment: thus, all titles not aligned with the theme are excluded. The third filter is performed according to the representativeness of the article (number of citations) through Google Scholar. We chose to take into account all articles with more than one citation. In the articles without citation, only articles published in 2020 were read. The next step is to read the abstracts, in order to verify the representativeness of the articles.

After applying all the filters, the articles were read. Table 3 presents the bank of articles collected on sustainable healthy cities, in the period from 2010 to 2020. The references of these articles were also used; that is why we used is an article from the year 2000.

These selected articles will be the basis for composing the indicators. However, other articles selected in the database will compose this research and support these indicators.

3.2 Procedures for Elaborating Indicators

To elaborate our indicators, the indicators cited in the literature were identified (articles presented in the previous topic), then tabulated in an electronic spreadsheet, and the most cited works gave indicators their names.

4 Results and Analysis

An alternative for monitoring sustainability is the use of performance indicators. Sustainability indicators are used in order to monitor sustainable development. Thus, these indicators are responsible for capturing information that will compose an informative diagnosis for decision-makers, guiding the development and monitoring of policies and strategies. Sustainable performance indicators are identified in the literature and classified into ten categories: social, economic, governance, energy, environment, transport, water, green spaces, air quality, and health.

Table 3 Bank of articles collected on healthy sustainable cities

Authors	Journal
Rotmans et al. [40]	*Environmental Impact Assessment Review. vol*
Rosales [39]	*Procedia Engineering*
Bao and Toivanen [5]	*Journal of Science and Technology Policy Management*
El Ghorb and Shalay [25]	*Alexandria Engineering Journal*
Anand et al. [3]	*Energy Procedia*
Taecharungroj et al. [50]	*Journal Of Place Management And Development*
Brilhante and Klaas [11]	*Sustainability*
Giles et al. [26]	*Health Policy*
Silva et al. [17]	Revista de Gestão—REGE
Su et al. [48]	*Ecological Indicators, 9:*
Deng et al. [18]	*Cities*
Sokolov et al. [45]	*Technological Forecasting and Social Change*
Alyami [2]	*IEEE Access*
Yang et al. [61]	*Resources Policy*
Ruan et al. [41]	*Cities*
Wang and Peng [60]	*Mathematics*
Steiniger et al. [47]	*Cities*
Li and Yi [32]	*Journal of Cleaner Production*
Kourtit t al. [30]	*Science of the Total Environment*
Jing and Wang [29]	*Journal of Cleaner Production*

Source Prepared by the Authors, 2020

Table 4 presents the Strategic Map with the indicators, the description of each indicator and the authors that support these indicators.

The indicators of healthy smart sustainable cities are now related to the COVID-19 pandemic instructions and the lessons we learned from it.

Now, the social, economic, governance, energy, environment, transport, water, green spaces, air quality, and health indicators will be analyzed individually.

4.1 Social Indicator

The first analyzed indicator is the social indicator, which aims primarily at social welfare through quality food, access to housing, population density, social equity, employment structure, life expectancy, cultural identity, general employment index,

Table 4 Strategic map of indicators

Indicators for healthy smart sustainable cities		
1	Indicator	Social Indicator
Description		Investments in the social welfare of the inhabitants of the city
Authors		[3, 11, 17, 18, 25–27, 29, 31–41, 45, 47, 48, 60, 61]
2	Indicator	Economic indicator
Description		Economic development and growth of a city
Authors		[3, 11, 17, 25, 27, 29, 30, 31–41, 45, 48, 61]
3	Indicator	Governance indicator
Description		Governance level of a city
Authors		[3, 12, 25, 45, 47, 50]
4	Indicator	Energy indicator
Description		Investments and valorization of issues related to the city's energy
Authors		[2, 3, 5, 11, 12, 18, 32, 41, 45, 49–50, 58]
5	Indicator	Environment indicator
Description		Investments and valorization of issues related to the environment in a city
Authors		[2, 3, 11, 17, 18, 29, 30, 31, 32, 33, 34, 35, 36, 37, 38, 39, 40, 41, 48]
6	Indicator	Transport indicator
Description		Investments and accessibility in the city transport
Authors		[1– 3, 11, 18, 27, 31, 33, 45, 49, 50, 58]
7	Indicator	Water indicator
Description		Investments and valorization of issues related to water management in the city
Authors		[2, 5, 11, 12, 26, 29, 32, 34, 40–50]
8	Indicator	Green spaces indicator
Description		Investments, valorization and access to green spaces in the city
Authors		[1, 18, 29, 31, 33, 34, 40, 45]
9	Indicator	Air quality indicator
Description		Monitoring and preservation of the city's air quality

(continued)

Table 4 (continued)

Indicators for healthy smart sustainable cities		
1	Indicator	Social Indicator
Authors		[1, 3, 25, 29, 30, 31,33, 34,40, 45]
10	Indicator	Health indicator
Description		Investments, valorization and access to health services in the city
Authors		[3, 5, 26, 27, 31, 39, 47]

urban vulnerability, internet access, assistance from municipal services, general consumer price index, income level, community projects, unemployment rates, and many other issues involving the social indicator that are of concern in the pandemic [3, 11, 17, 18, 25, 26, 27, 29, 31–41, 45, 47, 48 60, 61].

The social indicator includes results on topics such as quality of life sustainability, sustainable development, and the relationship between quality of life and sustainability. The social indicator assesses the population's quality of life and well-being levels and the realization of social and human rights. It also discusses access to services, goods, and opportunities.

The pandemic has not only brought about a collapse in the health system of many cities but has also increased unemployment rates, numbers of children away from school, domestic violence, and other consequences.

Many communities lack access to quality food, according to [59]. When handling food, it is necessary to wash your hands with soap and water and also sanitize food, especially if you eat without cooking. A city that aims for health, invests in a healthy diet, and assists the population in terms of handling and consumption, according to [26] should also invest in nutritional literacy.

Hygiene is one of the main things that people should pay attention to in the context of a pandemic, not only regarding food, but also clothes, hands, and masks. [59] Stresses that cloth do not need to be cleaned with a washing machine or dryer, but the use of detergent or soap is necessary.

Access to education was hampered by the pandemic. While several cities aim for a safe return, increases in virus transmission, and a lack of hospital capacity emergency measures such as lockdowns are necessary. Furthermore, according to UM (2021) the closure of schools, lack of support network, economic limitations, and even the death of parents by COVID-19, has increased the risk of child marriage for girls.

According to [53], the closure of schools increases the likelihood of permanent school dropout. For the reopening of schools, policies for access to health and social services will be necessary.

A return to school still requires strategic planning for the safety of students and teachers. According to [59] for the protection of oneself and others, it is necessary to keep the distance of at least one meter between people, use masks when being with other people and clean the masks properly; maintain hand hygiene, among other ways to avoid infection.

According to [60], in addition to quantifiable physical standards, a sustainable city needs to improve the quality of local life, so that this improvement is perceived by residents. Agenda 21 (1992) has as its goal a healthy life, so the WHO (1997) proposed the healthy city (HC) project, intending to promote sustainable and healthy urban development. According to WHO (1998), a healthy city aims to create and continuously improve the physical and social environment, allowing people to support each other in order to develop their potential.

Reference [40] proposes integrated city planning, aiming to improve several factors that make up a city with the efficient use of resources, facilitating the creation of sectoral policies. In this way, economic, socio-cultural, and ecological areas will be covered, in their sub-indicators such as: Resources/materials, work structure, transport infrastructure, demographic structure, knowledge structure, cultural heritage, quality and quantity of natural resources, and biodiversity.

Brito et al. [12] developed in their research a multiple criteria model aimed at assessing the sustainability of green cities using the multiple criteria decision approach (MCDA). The authors reached the following criteria: people, mobility, water, energy efficiency, biodiversity, waste, governance, and innovation.

4.2 Economic Indicator

The economic indicator corresponds to the economic development and growth of a city [3, 11, 17, 25, 27, 29, 30, 31–41, 45, 48, 61].

One possible economic indicator is the use of economic statistics, such as the unemployment rate, GDP, or inflation rates. They indicate how the economy is doing and predict how the economy will perform in future. For sustainable cities, economic indicators serve as a parameter for generating wealth and the ability to invest in sustainable actions.

The economic indicator involves the attractiveness of the market, economic development, income concentration, GDP per capita, contingent plan, the proportion of expenses with science and technology, comparison between the gain of urban families and the gain of rural families, growth rate of revenue regional, export rate, foreign investment, among other issues. Many of those have suffered the impacts of the pandemic.

According to [43], economic shutdowns had negative impacts on the urban economy. The consequences are diverse and involve other areas, including the already mentioned social issues. Also according to the authors, the pandemic impacted municipal tax revenues, population income, tourism, small and medium-sized companies, the food supply chain, migrant workers, and inequality in social and spatial distribution.

The impacts of COVID-19 are faced by social groups differently and disproportionately. Poorer people and in marginalized regions are more likely to suffer social and economic damage. Periods of recession increase unemployment, so designing and implementing programs aimed at post-pandemic recovery will be important [43].

4.3 Governance Indicator

The indicator related to governance is related to planning in sustainable cities, corruption, economy potential [3] (Brito et al. [12]), people's participation, but also, environmentally friend purchases [25]. Still, according to [47] governance in a healthy smart sustainable city must have citizen participation in the elections, the government's openness to requests for information, transparency, and municipal budget dependency. Furthermore,

Reference [50] mention the importance of governance. In a pandemic context, it is not difficult to reflect the value of governance, because, more than ever, governments and companies need a set of practices in order to improve management. There are so many issues to manage in a city, in times of pandemic, the problems become accentuated, where risk management becomes necessary.

According to [21] the conditions for a healthier, safer, and more resilient post-pandemic society must already be under construction, allowing for better preparation and alert for future risks. Still according to [21], (p 7) returning to the normal "situation before the pandemic means to maintain the same conditions of risks and vulnerabilities that caused the global disaster by COVID-19".

Some cities have shown a disconnect between the political sphere and collective participation in planning and decision-making. Thus, the democratization of the participation of the population helps to build governance, as well as the exercise of citizenship through the social and collective appropriation of the city.

4.4 Energy Indicator

The energy indicator is related to several sectors of a city and is of special importance in a pandemic context when considering the energy demand of a hospital. Energy includes the use of renewable energy, smart housing, smart automation [3, 5, 12, 45, 50]. Still, according to [2], a sustainable city must have the monitoring of emissions, reduction of the carbon footprint, measurement of the spent energy, renewable energy potential. Subadyo et al. [49] mention the potential of solar energy and hydroelectric power in cities. [48] Mention the measurement of energy consumption by GDP. Examples of technologies that can be used to generate clean energy are the generation of photovoltaic energy, hybrid wind energy systems; hybrid solar energy systems; bioenergy; and geothermal energy [28].

The concern with alternative forms of energy generation, especially those from renewable sources, is at the base of a sustainable city. Cities that are concerned with sustainable development have sought to reduce the use of fossil fuel/thermoelectric energy and carbon emissions from their energy matrix. Cities are discovering more efficient ways of management, improving the quality of life, and creating less impact on human health through renewable energies.

4.5 Environment Indicator

The environment indicator is broad and also involves several sectors in the city. The themes related to this indicator are energy-efficient technologies, clean transport, air pollution (including noise) [3], forestry area, wastewater capacity, use of solid waste [29], green spaces [1, 18, 29, 31, 33, 34, 40, 45].

The pandemic has had positive and negative impacts on the environment. Among the positive impacts are the decrease in air pollution, noise pollution, and greenhouse gas emissions [19].

The environment is also negatively impacted by the pandemic, as various wastes are improperly disposed of and become a threat. Reference [19] investigated the impacts of facial mask residues on the marine environment. According to the authors, many masks are made of petroleum-based polymers, which are not biodegradable.

Reference [19] highlights the importance of waste management with the sorting, collection, treatment, and safety protocol. Finally, the authors propose as an alternative to the problem of marine pollution by facial masks, the production of masks with nanotechnology or biomaterial.

References [4] and [6] also studied pollution on beaches, coasts, and seas by polymeric and antiviral textile waste. They emphasize that poor management represents an environmental problem, and therefore, waste management strategies must be studied.

As for noise pollution, a concern is its negative effects on physical and mental health since frequent exposure to noise above the recommended level causes serious damage to the body, such as heart problems, anxiety attacks, and gradual hearing loss.

Regarding the pandemic, there are several issues involved, because in lockdowns or social distancing policies, people travel less by plane, stay at home reduce the use of their own or public transport, creating a trend of reduction in pollution levels as well.

The main objective of a sustainable city is to avoid the depletion of the environment of its territory, thus guaranteeing its permanence for future generations. In this way, sustainable cities seek to reinvent themselves so that in future the next generations will have a planet more conducive to human life in order to guarantee the quality of life.

Reference [55] investigated the effects of the COVID-19 pandemic on environmental protection and legislation in Brazil. The research aimed to evaluate the main legislative actions, environmental fines, and deforestation. The results showed that 57 legislative acts during the current administration aimed to weaken environmental protection, half of which almost were created during the pandemic. The results also showed that during the pandemic there was a 72% reduction in environmental fines, although there was an increase in deforestation in the Amazon in the same period. Reference [55] also conclude that these actions can generate a loss of biodiversity, increase greenhouse gas emissions, and the likelihood of other outbreaks of zoonotic diseases.

4.6 Transport Indicator

The transport indicator corresponds to investments and accessibility in transporting a city [1–3, 11, 18, 27, 31, 33, 45, 49, 50, 58].

Sustainable transport is a form of strategic locomotion that aims to improve the lives of people in cities and to guarantee the right of citizens to come and go. Transport is a means that allows the population access to a set of needs that contribute to healthy and fulfilling lives, such as employment, markets, social interaction, and education. Examples of sustainable urban mobility are urban belts, exclusive routes for public transportation, cycle tracks, and cycle lanes, electric car-sharing network, and integration of different modes.

Reference [51] studied the impact of the pandemic on urban transport and air quality in Canada. According to the authors, due to the blockages caused by the pandemic, the demand and domestic consumption of fuel for transportation had a considerable decline. There were drops in the monthly consumption of transportation fuel in general as well, that is, diesel oil, kerosene-fuel for aircraft, and gasoline for engines. In May 2020, gasoline consumption dropped by 49.8% and diesel oil dropped by 28.4% compared to 2019. According to [51] "in week 12, the level of traffic congestion decreased by 69% and 75% in Toronto and Montreal, respectively, compared to 2019".

4.7 Water Indicator

The water indicator corresponds to the investment and valorization of water issues in the city [2, 5, 11, 12, 26,29, 32, 34, 45, 40–50].

Some actions are important to overcome water crises and ensure access to drinking water for the population, such as water reuse, desalination, reversing of pollution, river crossing, rainwater harvesting, cleaning streams, basic sanitation. In 2017, about two billion people did not have access to drinking water (WHO, 2017). For this reason, the search for solutions for the appropriate use of water resources becomes even more important.

4.8 Green Spaces Indicator

The green spaces indicator corresponds to investments, valorization, and access to green spaces in the city [1, 18, 29, 31, 33, 34, 40, 45].

The use of green spaces within the limits of a city aims to improve the urban environment. In this way, green spaces improve air quality, reduces the effects of heating paved surfaces, promote soil drainage, and, consequently, replenishes groundwater reserves, lakes, and rivers.

Reference [37] researched the benefits of contact with green spaces during COVID-19 lockdowns for mental health. The authors concluded that contact with nature mitigates the negative effect of the lockdowns on mental health. Most people understand that contact with nature helps to deal with restrictive measures in a better way. The study also demonstrates that access to green outdoor spaces with views of nature is associated with more positive emotions.

Still concerning access to green spaces during the pandemic, the research by [54] took into account the effects of the pandemic on the use and perceptions of urban green space. The results showed that the residents surveyed need accessible urban green spaces so that they can exercise, relax, take the dog for a walk and observe nature.

4.9 Air Quality Indicator

The air quality indicator corresponds to the monitoring and preservation of the city's air quality [1, 3, 25, 29, 30, 31, 33, 34, 40, 45].

Air quality management in cities aims to ensure that development takes place in a sustainable and environmentally safe manner. To this end, cities should promote actions that encourage the prevention, combat, and reduction of pollutant gas emissions and the effects of the degradation of the atmospheric environment. Air pollutants are gases and solid particles resulting from human activities and natural phenomena dispersed in atmospheric air. The main sources of pollutant emissions are motor vehicles, trains, airplanes, marine vessels, industrial chimneys, and power generation (thermoelectric). These sources must be countered by city decisions in order to improve air quality.

Reference [51] studied the impact of the pandemic on urban transport and air quality in Canada. According to a study carried out in Canada between 2018 and 2020, the transport and mobile equipment sector was responsible for generating 56% of total CO emissions. However, according to [51] "the level of CO concentration in six cities has decreased since March 2020". The authors also conclude that the quality of urban air across the country has greatly improved during the pandemic period, but it is still unclear how the recovery will be".

4.9.1 Health Indicator

The health indicator corresponds to investments, valuation and access to health services in the city [3, 5, 26, 27, 31, 39, 47].

Today, more than half of the world's population now lives in urban centers. Therefore, cities should be concerned with promoting the search for health and quality of life, especially in the post-COVID-19 context. Parameters of health and well-being applied to urban development were created through environmental certifications Well and Fitwel.

These stamps use sustainability indicators to encourage the development of projects, practices that promote health in urban spaces. Actions such as the creation of urban farms, contemplative spaces for mental restoration, publicly accessible drinking fountains, cycling infrastructure, and encouraging walking are the basis for the health and well-being of the population. Healthy cities seek to think of urban health as a resource for the life and happiness of its inhabitants.

Around the world, governments and health professionals are experiencing severe stress. Patient care is changed according to the need of the moment. Many health professionals are also being infected and away from work. According to [24], the emotional suffering or exhaustion of health professionals was rated as high, especially for nurses. This problem is also associated with the lack of personnel or the lack of resources and communication, which is often deficient, on the part of supervisors.

Also according to [24] "the recovery from the peak pandemic situation will be better if people feel safe and connected to others during the crisis, if they have social, physical, and emotional support and feel they can help themselves and their community".

Reference [46] research focuses on food security in cities during periods of the pandemic. According to the authors, the global food supply chain operates in a balance between consumption, production, and stock. However, minor disturbances can cause instability, such as a pandemic.

The COVID-19 pandemic demonstrated how an infectious disease can lead to a food crisis, because many places have closed their borders, imposed lockdowns, and interrupted trade, resulting in a shortage of labor. These actions have caused disruptions along the food supply chain. The consequence of this process is limited access to food, especially for vulnerable groups [46].

Still, concerning food, the pandemic can modify consumer behavior, as they can buy more than they usually do, aiming at a private stock as a way to prevent scarcity, causing an increase in food prices, destabilizing the food system. Cities are even more vulnerable than rural areas, as cities depend on food imports [46].

Reference [35] presents a counterpoint concerning this problem of food scarcity during a pandemic, as they state that although the news was initially about empty shelves and the closing of schools, hotels, and restaurants have led many producers to destroy unused food products.

5   Final Considerations

The research proposed to make a parallel between the needs imposed by COVID-19 and the characteristics of healthy smart sustainable cities, analyzing how the concept of sustainable cities can contribute in a pandemic context. Sustainable performance indicators are identified in the literature and classified into ten categories: social, economic, governance, energy, environment, transport, water, green spaces, air quality, and health. After choosing these indicators, we have identified the main characteristics of each, relating them both to the concept of healthy smart sustainable cities and research related to the COVID-19 pandemic.

The international emergency related to COVID-19, declared on January 30, 2020, by the World Health Organization (WHO), has turned the attention of the scientific community once again to the field of global health and sustainable development [56].

Buckeridge and Philippi Junior (2020) add that the COVID-19 pandemic revealed the virtues and deficiencies of cities in facing one of the strongest extreme events of the twenty-first century. On the one hand, the strength of science has helped to tackle the pandemic, advising on health issues at the peak of the disease. On the other hand, the deficiencies in public policies that are problems of the past demanded their resolution, revealing the most perverse face of the existing inequality in the city, the vulnerability to extreme events.

To complement the discussion, the authors Buckeridge and Philippi Junior [14] explain the relations between this pandemic and sustainable cities, making an analogy with ecology, in which the city can be seen as a urbsystem with a primary and a secondary structure, latter housing networks of services that determine the quality of life in the city. The generation of public policies is the main "physiological" mechanism by which cities can become more or less vulnerable to extreme events, such as climate change and pandemics.

The literature also shows that the COVID-19 pandemic highlights the importance of urban sustainability. It shows that sustainability can only be achieved through intelligence that always considers the application of public policies based, this can be done through broader communication with the scientific community.

Reference [15] highlighted the impacts of the pandemic on the social vulnerability that can be perceived as the resilience of communities when confronted by external factors that are stressful to health, such as natural or human-caused disasters, or disease epidemics. Reducing social vulnerability can decrease both human suffering and economic losses.

Thus, one of the measures applied and to seek the sustainability of cities at this time of COVID-19 pandemic, according to the recommendation of the Ministry of Health, in compliance with the National Contingency Plan for Human Infection by the novel coronavirus, should include social isolation, and other procedures such as strict hand hygiene and respiratory etiquette, as well as avoiding crowds, maintaining a minimum distance of 1.5 m between people in outdoor environments, and above all staying at home—except for those who work essential activities, such as supermarkets, pharmacies, health units, and others.

However, the authors [15] indicate emphasizes that in order to meet such measures, it is necessary to reflect on which lives in communities that do not have adequate infrastructure and basic sanitation, thus presenting difficulties about personal hygiene and the environment. Many families live in narrow rooms, in conditions that do not favor isolation if someone becomes infected with the coronavirus. The main source of family income is often derived from informal and face-to-face activities, making it impossible to work from home, among many other problems.

Acknowledgements This study was conducted by the Centre for Sustainable Development (Greens), from the University of Southern Santa Catarina (Unisul) and Ânima Institute—AI, in the context of the project BRIDGE—Building Resilience in a Dynamic Global Economy: Complexity across scales in the Brazilian Food-Water-Energy Nexus; funded by the Newton Fund, Fundação de Amparo à Pesquisa e Inovação do Estado de Santa Catarina (FAPESC), Coordenação de Aperfeiçoamento de Pessoal de Nível superior (CAPES), National Council for Scientific and Technological Development (CNPq) and the Research Councils United Kingdom (RCUK).

References

1. Mueller N, Rojas-Rueda D, Khreis H, Cirach M, Andrés D, Ballester J, Bartoll X, Daher C, Deluca A, Echave C, Milà C (2020) Changing the urban design of cities for health: The superblock model. Env Int 134:105132 https://doi.org/10.1016/j.envint.2019.105132
2. Alyami SH (2019) Opportunities and Challenges of Embracing Green City Principles in Saudi Arabia Future Cities. IEEE Access 7:178584–178595. https://doi.org/10.1109/access.2019.2959026
3. Anand A, Rufuss DDW, Rajkumar V, Suganthi L (2017) Evaluation of sustainability indicators in smart cities for India using MCDM approach. Energy Procedia 141:211–215. https://doi.org/10.1016/j.egypro.2017.11.094
4. Ardusso M, Forero-López A, Buzzi N, Spetter C, Fernández-Severini M (2021) COVID-19 pandemic repercussions on plastic and antiviral polymeric textile causing pollution on beaches and coasts of South America. Sci Total Environ 763:144365. https://doi.org/10.1016/j.scitotenv.2020.144365
5. Bao S, Toivonen M (2014) The specificities and practical applications of Chinese eco-cities. J Sci Technol Policy Manage 5(2):162–176. https://doi.org/10.1108/JSTPM-05-2014-0020
6. Benson N, Bassey D, Palanisami T (2021) COVID pollution: impact of COVID-19 pandemic on global plastic waste footprint. Heliyon 7(2):e06343. https://doi.org/10.1016/j.heliyon.2021.e06343
7. Bento, SC, Conti DM, Baptista, RMB, Ghobril, CN (2018). As novas diretrizes e a importância do planejamento urbano para o desenvolvimento de cidades sustentáveis. Revista de Gestão Ambiental e Sustentabilidade 7(3).
8. Bibri S, Krogstie J (2017) Smart sustainable cities of the future: An extensive interdisciplinary literature review. Sustain Cities Soc 31:183–212. https://doi.org/10.1016/j.scs.2017.02.016
9. Boareto R (2008) A política de mobilidade urbana e a construção de cidades sustentáveis. Revista dos Transportes Públicos – ANTP
10. Brazilian Ministry of Health (2021) https://coronavirus.saude.gov.br/sobre-a-doenca. Retrieved February 2021\
11. Brilhante O, Klaas J (2018) Green city concept and a method to measure green city performance over time applied to fifty cities globally: influence of GDP, population size energy efficiency. Sustainability 10(6):23. https://doi.org/10.3390/su10062031
12. Brito VT, Ferreira FA, Pérez-Gladish B, Govindan K, Meidutė-Kavaliauskienė I (2019) Developing a green city assessment system using cognitive maps and the Choquet Integral. J Cleaner Production 218:486–497. https://doi.org/10.1016/j.jclepro.2019.01.060
13. Brundtland G (1987) Report of the World Commission on Environment and development: our common future: United Nations General Assembly document A/42/427
14. Buckeridge, MS, Philippi Jr A (2020) Ciência e políticas públicas nas cidades: revelações da pandemia da Covid-19. Estudos Avançados 34(99)
15. Christoffel MM, Gomes ALM, Souza TV, Ciuffo LL (2020) A (in)visibilidade da criança em vulnerabilidade social e o impacto do novo coronavírus (COVID19). Revista Brasileira de Enfermagem 73(supl.2)

16. Couto CFV, Medeiros GD, Alves MFP, Dias C, Braga IYLG, Andrade NP (2020) A pandemia da covid-19 e os impactos para a mobilidade urbana. ANPET.
17. Da Silva CA., Dos Santos EA, Maier SM, Da Rosa FS (2019) Urban resilience and sustainable development policies. Revista de Gestão 27(1):61–78. https://doi.org/10.1108/REGE-12-2018-0117
18. Deng W, Peng Z, Tang Y-T (2019) A quick assessment method to evaluate sustainability of urban built environment: case studies of four large-sized Chinese cities. Cities 89:57–69. https://doi.org/10.1016/j.cities.2019.01.028
19. Dharmaraj S, Ashokkumar V, Hariharan S, Manibharathi A, Show P, Chong C, Ngamcharuss-rivichai C (2021) The COVID-19 pandemic face mask waste: a blooming threat to the marine environment. Chemosphere 272:129601. https://doi.org/10.1016/j.chemosphere.2021.129601
20. Fabris JF, Bernardy RJ, Sehmem S, Piekas AAS (2019) Cidades sustentáveis: caminhos e possibilidades. Int J Prof Bus Rev
21. Freitas, CM (2020) A gestão de riscos e governança na pandemia por covid-19 no brasil: análise dos decretos estaduais no primeiro mês - relatório técnico e sumário executivo: Cepedes | Ensp Centro de Estudos e Pesquisas em Emergências e Desastres em Saúde
22. GIFE—Grupo de Institutos Fundações e Empresas (2021) 21 oportunidades para a atuação do investimento social privado no desenvolvimento sustentável das cidades em 2021. https://gife.org.br/especial-redegife-21-oportunidades-para-a-atuacao-do-investimento-social-privado-no-desenvolvimento-sustentavel-das-cidades-em-2021/, Retrieved February 2021
23. GPS. Guia GPS Gestão Pública Sustentável (2021) www.cidadessustentaveis.org.br/gps. Acesso fevereiro de 2021. REIS, José. Palavras para lá da pandemia: cem lados de uma crise. FCT – Fundação para a Ciência e a Tecnologia no âmbito do projeto UIDB/50012/2020
24. Garros D, Austin W, Dodek P (2020) How can I survive this? Chest. https://doi.org/10.1016/j.chest.2020.11.012
25. El Ghorab HK, Shalaby HA (2016) Eco and Green cities as new approaches for planning and developing cities in Egypt. Alex Eng J 55(1):495–503. https://doi.org/10.1016/j.aej.2015.12.018
26. Giles B, Lowe M, Arundel J (2019) Achieving the SDGs: Evaluating indicators to be used to benchmark and monitor progress towards creating healthy and sustainable cities. Health Policy. https://doi.org/10.1016/j.healthpol.2019.03.001
27. He X, Lin M, Chen TL, Lue B, Tseng PC, Cao W, Chiang PC (2020) Implementation plan for low-carbon resilient city towards sustainable development goals: Challenges and perspectives. Aerosol and Air Qual Res. https://doi.org/10.4209/aaqr.2019.11.0568
28. Hua C, Chen J, Wan Z, Xu L, Bai Y, Zheng T, Fei Y (2020) Evaluation and governance of green development practice of port: A sea port case of China. J Clean Prod 249:119434. https://doi.org/10.1016/j.jclepro.2019.119434
29. Jing Z, Wang J (2020) Sustainable development evaluation of the society–economy–environment in a resource-based city of China: A complex network approach. J Cleaner Production 263:121510. https://doi.org/10.1016/j.jclepro.2020.121510
30. Kourtit K, Nijkamp P, Suzuki S (2020) Are global cities sustainability champions? A double delinking analysis of environmental performance of urban agglomerations. Sci Total Environ 709:134963. https://doi.org/10.1016/j.scitotenv.2019.134963
31. Langellier BA, Kuhlberg JA, Ballard EA, Slesinski SC, Stankov I, Gouveia N, Meisel JD, Kroker-Lobos MF, Sarmiento OL, Caiaffa WT, Roux AV (2019) Using community-based system dynamics modeling to understand the complex systems that influence health in cities: The SALURBAL study. Health & Place 60:102215. https://doi.org/10.1016/j.healthplace.2019.102215
32. Li W, Yi P (2020) Assessment of city sustainability—Coupling coordinated development among economy, society and environment. J Clean Prod 256:120453. https://doi.org/10.1016/j.jclepro.2020.120453
33. Lowe M, Arundel J, Hooper P, Rozek J, Higgs C, Roberts R, Giles-Corti B (2020) Liveability aspirations and realities: Implementation of urban policies designed to create healthy cities in Australia. Social Sci Med 245:112713. https://doi.org/10.1016/j.socscimed.2019.112713

34. Meerow S (2020) The politics of multifunctional green infrastructure planning in New York City. Cities 100:102621. https://doi.org/10.1016/j.cities.2020.102621
35. Meyer B, Prescott B, Sheng X (2021) The impact of the COVID-19 pandemic on business expectations. Int J Forecast. https://doi.org/10.1016/j.ijforecast.2021.02.009
36. Ministério da Saúde do Brasil. https://coronavirus.saude.gov.br/sobre-a-doenca. Retrieved 9 March 2021
37. Pouso S, Borja Á, Fleming L, Gómez- E, White M, Uyarra M (2021) Contact with blue-green spaces during the COVID-19 pandemic lockdown beneficial for mental health. Sci Total Environ 756:143984. https://doi.org/10.1016/j.scitotenv.2020.143984
38. Reis J (2020) Palavras para lá da pandemia: cem lados de uma crise. Financiado por Fundos FEDER através do Programa Operacional Factores de Competitividade – COMPETE e por Fundos Nacionais através da FCT – Fundação para a Ciência e a Tecnologia no âmbito do projeto UIDB/50012/2020
39. Rosales N (2011) Towards the modeling of sustainability into urban planning: using indicators to build sustainable cities. Procedia Eng 21:641–647. https://doi.org/10.1016/j.proeng.2011.11.2060
40. Rotmans J, Asselt MV, Vellinga P (2000) An integrated planning tool for sustainable cities. Environ Impact Assess Rev 20:265–276
41. Ruan F, Yan L, Wang D (2020) The complexity for the resource-based cities in China on creating sustainable development. Cities 97:102571. https://doi.org/10.1016/j.cities.2019.102571
42. Senhoras EM, Gomes ML (2020) COVID-19 nos municípios de roraima. BOCA. Boletim de Conjuntura. Ano II, Volume 3, N° 9, Boa Vista. http://dx.doi.org/10.5281/zenodo.4036180
43. Sharifi A, Khavarian-Garmsir A (2020) The COVID-19 pandemic: Impacts on cities and major lessons for urban planning, design, and management. Sci Total Environ 749:142391. https://doi.org/10.1016/j.scitotenv.2020.142391
44. Silva B, Khan M, Han K (2018) Towards sustainable smart cities: A review of trends, architectures, components, and open challenges in smart cities. Sustain Cities Soc 38:697–713. https://doi.org/10.1016/j.scs.2018.01.053
45. Sokolov A, Veselitskaya N, Carabias V, Yildirim O (2019) Scenario-based identification of key factors for smart cities development policies. Technol Forecast Soc Chang 148:119729. https://doi.org/10.1016/j.techfore.2019.119729
46. Song S, Goh J, Tan H (2021) Is food security an illusion for cities? A system dynamics approach to assess disturbance in the urban food supply chain during pandemics. Agric Syst 189:103045. https://doi.org/10.1016/j.agsy.2020.103045
47. Steiniger S, Wagemann E, de la Barrera F, Molinos-Senante M, Villegas R, de la Fuente H, Vives A, Arce G, Herrera JC, Carrasco JA, Pastén PA, Barton JR (2020) Localising urban sustainability indicators: The CEDEUS indicator set, and lessons from an expert-driven process. Cities 101:102683. https://doi.org/10.1016/j.cities.2020.102683
48. Su M, Xie H, Yue W, Zhang L, Yang Z, Chen S (2019) Urban ecosystem health evaluation for typical Chinese cities along the Belt and Road. Ecol Ind 101:572–582. https://doi.org/10.1016/j.ecolind.2019.01.070
49. Subadyo AT, Tutuko P, Jati RMB (2019) Implementation analysis of green city concept in malang - Indonesia. Int Review for Spatial Plann Sustainable Dev 7(2):36–52. https://doi.org/10.14246/irspsd.7.2_36
50. Taecharungroj V, Tachapattaworakul T, Rattanapan C (2018) The place sustainability scale: measuring residents' perceptions of the sustainability of a town. J Place Manage Dev 11(4):370–390. https://doi.org/10.1108/jpmd-04-2017-0037
51. Tian X, An C, Chen Z, Tian Z (2021) Assessing the impact of COVID-19 pandemic on urban transportation and air quality in Canada. Sci Total Environ 765:144270. https://doi.org/10.1016/j.scitotenv.2020.144270
52. UN (2021) Pandemia pode levar mais de 10 milhões de meninas a casar cedo. https://news.un.org/pt/story/2021/03/1743772. Retrieved 10 March 2021
53. UNICEF (2021) https://secure.unicef.org.br/Default.aspx?origem=drtggl&gclid=Cj0KCQiA-aGCBhCwARIsAHDl5x-DrxdBZNxDbOmCglKRTncDLWuclhQ6ofktSK94zrOAYqfaKZ0IWHkaApFREALw_wcB. Retrieved 10 March 2021

54. Ugolini F, Massetti L, Calaza-Martínez P, Cariñanos P, Dobbs C, Ostoić S et al (2020) Effects of the COVID-19 pandemic on the use and perceptions of urban green space: An international exploratory study. Urban Forestry Urban Green 56:126888. https://doi.org/10.1016/j.ufug.2020.126888

55. Vale M, Berenguer E, Argollo M, Viveiros E, Pugliese L, Portela R (2021) The COVID-19 pandemic as an opportunity to weaken environmental protection in Brazil. Biol Cons 255:108994. https://doi.org/10.1016/j.biocon.2021.108994

56. Ventura, DFLV, Ribeiro H, Giulio Gabriela Marques Di, Jaime PC, Numes J, Bogus CM, Antunes JLF, Waldman EA (2020) Desafios da pandemia de COVID-19: por uma agenda brasileira de pesquisa em saúde global e sustentabilidade. Cad. Saúde Pública

57. Vieira JSR (2012) Cidades sustentáveis. Revista de Direito da Cidade 4(2)

58. Vukovic N, Rzhavtsev A, Shmyrev V (2019) Smart city: The case study of Saint-Peterburg 2019. Int Rev (1-2) 15–20. 10.5937/intrev1901015V

59. WHO (2021) Coronavirus disease (COVID-19) advice for the public. https://www.who.int/emergencies/diseases/novel-coronavirus-2019/advice-for-public Retrieved 9 March 2021

60. Wang WM, Peng HH (2020) A fuzzy multi-criteria evaluation framework for urban sustainable development. Mathematics 8(3):330. https://doi.org/10.3390/math8030330

61. Yang Y, Guo H, Chen L, Liu X, Gu M, Ke X (2019) Regional analysis of the green development level differences in Chinese mineral resource-based cities. Resour Policy 61:261–272. https://doi.org/10.1016/j.resourpol.2019.02.003

62. Yigitcanlar T, Han H, Kamruzzaman M, Ioppolo G, Sabatini-Marques J (2019) The making of smart cities: Are Songdo, Masdar, Amsterdam, San Francisco and Brisbane the best we could build? Land Use Policy 88:104187. https://doi.org/10.1016/j.landusepol.2019.104187

Climate Change and COVID-19: Crisis Within Crises for Eradication of Poverty in Bangladesh

Shamima Ferdousi Sifa, Rukhsar Sultana, and Md. Bodrud-Doza

Abstract Bangladesh has made tremendous progress in poverty reduction over the past years and is now battling against the COVID-19 pandemic and climate change. The combined impacts of COVID-19 and climate change have derailed the progress of poverty reduction by creating a pool of "new poor" that is making poor people poorer and leading extreme poor into destitute. Climate change, COVID-19 and poverty interact and are interlinked with each other to increase the vulnerability of the population creating a long-lasting socio-economic crisis, food insecurity, inequalities, gender-based violence and discriminations against poor, marginalized and vulnerable groups. In order to address the current and impending challenges in the progress towards poverty reduction, a comprehensive and holistic approach is needed to implement a pro-poor strategy addressing climate change resilience, green economic recovery and sustainable development that will promote climate adaptive agricultural diversification, infrastructural development for safeguarding the vulnerable households, health and social security, creating employment opportunities, labour-intensive and export-orientated manufacturing-related inclusive green growth, modern service sector and overseas employment for transformative adaptation, youth development and entrepreneurship. Therefore, this chapter explores the issues related to the COVID-19 pandemic and climate change vulnerability that contributes to poverty, affect the path for eradication of poverty and policy response undertaken by the country to address these crises.

S. F. Sifa
Department of Disaster Science and Management, University of Dhaka, Dhaka 1000, Bangladesh

R. Sultana
International Centre for Climate Change and Development (ICCCAD), Independent University Bangladesh (IUB), Dhaka 1229, Bangladesh

Md. Bodrud-Doza (✉)
Climate Change Programme, BRAC, Dhaka 1212, Bangladesh
e-mail: bodrud.d@brac.net

Graphical Abstract

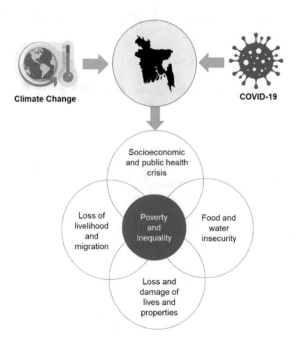

Keywords Sustainable development goals (SDGs) · COVID-19 pandemic ·
Climate change · Poverty · Bangladesh

1 Introduction

The current COVID-19 health crises amid the climate crisis have spread across the globe, upending the lives of millions of people and significantly affecting the global economy. As the number of deaths from COVID-19 is soaring up daily, the crisis overwhelmed the health systems globally, resulting in the closure of the educational institutes, business centres, factories, and disruption of the global supply system and value chains in many countries [1] which has a severe repercussion in the livelihood of the global workforce. United Nations (2020) expects that the COVID-19 crisis will push tens of millions of people into extreme poverty and hunger as the world struggles to contain the virus by undertaking several aggressive measures like nationwide shutdown, maintenance of social distancing, etc. which in turn have serious implications on both the economy and the population of the country. The Asian Development Bank (ADB) projects a global loss ranging from $77 billion to $347 billion, whereas United Nations Conference on Trade and Development (UNCTD) estimates that the loss incurred globally due to the pandemic will be $1 trillion [2]. Additionally, the International Labor Organization (ILO) projected around 25 million people will lose jobs and workers could lose some $3.4 trillion income by the end of the year 2020 globally [2]. As the world is entering into a recessionary period according to IMF [3], it has exposed the underlying weaknesses in the international system which will severely impact the existing infrastructure and services and SDGs in long term [1]. The economic impact of the crisis is estimated at $2 trillion and the other socio-economic impacts will be so profound that it is assumed to erase some of the laudable achievements of SDGs [3]. Other possibilities are that post-pandemic recoveries will result in the re-allocation of resources to meet the immediate priorities, which in turn will slow down the progress of some goals [4]. The challenges continue as several natural disasters hit the climate-vulnerable regions amid the pandemic. Events such as Cyclone Amphan in Bangladesh, Tropical Cyclone Harold in Vanuatu, erratic rainfall, flood and locusts outbreak in East Africa, climate change-induced food insecurity and malnutrition in Africa's volatile Sahel region and Yemen, respectively, forced the vulnerable communities to cope with the compounding crises concurrently, eliminating the progress of sustainable development goals (SDGs) [5, 6]. Therefore, the COVID-19 pandemic acts as a shock to the climate-vulnerable regions, which are already dealing with the adverse impacts of climate change, facing difficulties to adapt to the climate-related risks and other multiple crises.

It is challenging for the least developed countries (LDC) like Bangladesh to battle against the pandemic, climate change-induced extreme events, economic fallout and poverty [7]. Bangladesh is one of the most climate-vulnerable countries due to its geographical, geomorphological and topographical characteristics which along with high population density, poverty, unplanned development and anthropogenic activities, low economic and technological capacity increases the vulnerability of the population living in the hazard-prone area [8]. The country has been ranked as the seventh most affected country in the Global Climate Risk Index (CRI) in 2019, due to extreme weather events in the last 20 years, which alone claimed 407 people's life in 2018 [9, 10]. In addition, over 1% of the country's GDP is lost annually for disasters, and climate change impacts have exacerbated the situation further. To add more to it, poverty is a serious concern for the country as about 21.8% of the population of the country lives under poverty [11]. International Labour Organization (ILO) identified that Bangladesh had 2.9 million people unemployed and projected the number to go up by 3 million [12]. Even among the employed, 9.2% of the employed population are living below $1.90 PPP (Purchasing Power Parity) a day in 2019 [11]. About 70% of the people in the country used to live from hand-to-mouth before the COVID-19 crisis which is also now about to rise due to the pandemic [13, 14]. Asian Development Bank predicted Bangladesh to be overwhelmed by this pandemic, with the approximate loss of $3 billion in GDP (i.e. 1.10% of total decline) and nine million job losses [14]. As the COVID-19 crisis started stressing the existing economic system, it is mostly affecting marginalized and vulnerable groups of the country whose livelihoods are already under threat due to high exposure to natural and climate change-induced disasters [15].

However, Bangladesh as a UN member state has made considerable progress with the sustainable development goals. The steady implementation of the goals in the national agendas and incorporating them into the government's 7th and 8th five-year plans is helping it to achieve the remaining targets. Yet when the COVID-19 has made its course in the country, the progress has undergone a major setback. COVID-19 pandemic in a climate-vulnerable country with poverty on the rise is threatening developing countries like Bangladesh who are trying to break free of the poverty cycle. However, it also provides opportunities to adopt a sustainable recovery path with concentrated efforts to deal with loss and damage from both pandemic and climate change and increased investment in pro-poor and inclusive growth, green economy, inclusive societal actions and conservation of the environment [1, 16, 17]. The progress the country made from severe structural impediments to sustainable development is outstanding and noteworthy. Sustainable policy measures and economic strategies are designed in such a way to achieve the vision where poverty will be eliminated through rapid inclusive growth and transformational change. The chapter thereby discusses how COVID-19 pandemic and climate change vulnerability contributes to poverty, affect the path for eradication of poverty and policy response undertaken by the country to address these crises.

2   Poverty Status and Progress in Poverty Reduction

Over the years, since 2000, Bangladesh has made tremendous progress in poverty reduction, which is an inspiration for others [18]. According to [19], Bangladesh is a role model for reducing poverty with sustained economic growth which will further advance development and poverty alleviation. [18] describes that the progress was sustained by 6% + growth over the decade, which later in 2016 increased to 7.3%. Rapid GDP growth and progress of other development indicators have reduced the poverty rate greatly. In 1970, the poverty incidence was about 80% which declined to 24.3% in 2016 and has been estimated to be further reduced to 20.5% in 2019 [18]. Despite the accelerated growth of GDP, the rate of poverty reduction was at a slower pace in 1990–2000 but was faster during 2000–2010 compared to the slow pace of reduction during 2010–2016 due to a decrease in growth elasticity of poverty (GEP). This is because of the worsening of the inequality as the distribution of income becomes unequal which is linked to poverty and thereby slows down the poverty reduction. Between the years 2010–2016, the average income of people had increased, and the number of people below the hard-core poverty line had reduced, but at the same time, the country experiences an increase in income disparity amongst the rich and the poor. The income inequality for this period increased from 0.45 to 0.48% [20]. National Extreme poverty had lowered from 12.9% (2016) to 9.4% (2018) [21]. This had further increased the income disparity of the bottom 5% to the top 5% of the population [20]. Inequality is a major concern to address the sustainable development goal and this can be a major barrier in the country's pathway to eradicate poverty (SDG 1). The trends in coverage of Social Safety Net Programmes had also experienced a major push from 13.06% in 2005 to 58.1% in 2019 [21]. Of these, the majority of the social safety net programmes are active in the rural part of Bangladesh, which is one of the reasons why poverty reduction in urban areas is lower than in the rural region. It has been observed that the living standards of the population improved as per capita GDP growth almost doubled as it increases to 5.90% per annum in 2010 from 3.05% per annum in 1990. Therefore, the income Gini coefficient (a standard measure of income inequality) and Palmer ratio (ratio of income shares of the top 10% and bottom 40% of the population) are used to define the growth inequality nexus, shows that both are increasing over the time as shown in Fig. 1, meaning that the gap between high- and low-income groups is on the rise [18]. This stagnation of economic growth might affect the achievement of sustainable development goals (SDG-1). Therefore, the government took redistributive fiscal policy to address income inequality and broadens the reach of the poor to formal services to ensure inclusivity.

On the other hand, essential services of health, education and social protection do have a tremendous impact on the SDG 1, which only makes SDG Goal 3 (Good health and Wellbeing), Goal 4 (Quality Education) and Goal 10 (Reduced Inequalities) as well as SDG Goal 8 (Decent Work and Economic Growth) more relevant to SDG 1. The proportion of the Bangladesh government's funding in these sectors has increased. In FY 2015, the government had allocated 4.81% for health, 12.82% for education and 12.72% for social protection. Later in the Fiscal Year 2016–2017, budget allocated to health, education, and social protection increased to 6.53%, 14.42% and 15.25%, respectively [22]. This altogether has contributed to reduction in the Multidimensional Poverty Index (MPI) which decreases from 0.292 in 2007 to 0.18 in 2019. An important issue to consider is that education is an important indicator of social inclusion and intergenerational mobility [20]. Therefore, it is an integral issue in achieving SDG 1. Furthermore, a lack of coherent policies and interest towards the disadvantaged and marginalized people of the country has been a concern for SDG 10 (targets Reduced Inequalities) which has had effected the achievement of SDG 1 [20]. The achievement of SDG 1 is shown below in Fig. 2 [23].

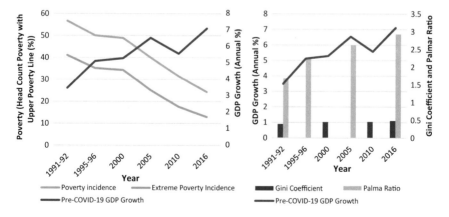

Fig. 1 Reduction of poverty, economic growth and income inequality nexus

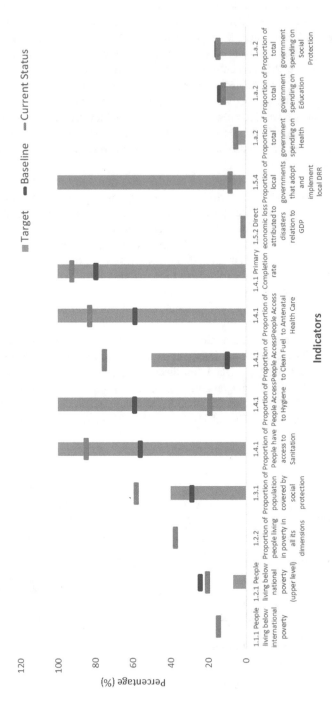

Fig. 2 Progress of SDG 1 (No Poverty) till 2020 has been shown where the grey bars represents targets and the blue and orange line indicates baseline and present conditions, respectively. It shows that targets has been achieved in many cases (e.g. 1.3.1, 1.4.1) but in some cases (e.g. 1.5.4, 1.a.2, etc.) the present conditions are same as baseline

However, there are many instances that the country has not only made progress in terms of rising per capita income but has also contributed to higher life expectancy, an increase of literacy rates and per capita food production, greater economic diversification, women's economic empowerment and gender mainstreaming [24]. In 2015, Bangladesh reached the lower-middle-income country status, and later in 2018, it reaches a new milestone by meeting all the three criteria to be eligible to graduate from the United Nation's Least Developed Countries (LDC) list by 2026 [25, 26]. The per capita gross national income (GNI) of the country was higher than the threshold in 2019 ($GNI_{Bangladesh}$:$1827 > $GNI_{Threshold}$:$1222) along with the other Human Asset Index (HAI) ($HAI_{Bangladesh}$:74.5 > $HAI_{Threshold}$:66) criteria and economic vulnerability index ($EVI_{Bangladesh}$:27.3 < $EVI_{Threshold}$:32) which needs to be less than the threshold value to meet the criteria [25]. Bangladesh is now on its way towards becoming a higher-middle-income country by 2031 and then will progress to a developed country by 2041 [27]. The graduation would have occurred before if the development had not been set back by the COVID-19 pandemic. In order to achieve the vision, the government is taking inclusive and aggressive strategies to make sure that the progress in poverty reduction is fully consistent with the SDG target of eliminating extreme poverty by 2030.

3 Impact of Climate Change in Poverty Reduction

Climate change and poverty interact are interlinked with each other in such a way that they both increase the vulnerability of the population [28, 29]. Climate change, directly and indirectly, contributes to the increase of vulnerability of the poor and individuals with inadequate access to assets, knowledge and adaptation information, employment disruption, fewer alternative livelihood options and exposure to high climate extremes, which perpetuate poverty. On the other hand, poverty increases the susceptibility of individuals towards climatic shocks and stresses [30, 31]. References [32] and [33] discuss how climate change interacts with poverty in a complex way where the adverse impacts of climate change on vulnerable groups are disproportionately distributed not only because they are poor but also because of their social vulnerability due to its difference in asset quantity, degree of social exclusion, existing inequalities, etc. This heterogeneity thereby increases the likelihood of being exposed to climatic extremes [34]. The adverse consequences also depend on the severity, timing, and other dimensions of both factors. Figure 3 shows a thematic diagram where the complex interaction of climate change and poverty is described. Reference [35] discusses that due to global warming and the rise of temperature by 1.5 °C will have an adverse impact as many would lose their livelihood option, loss of income, an increase in food price, etc. which will result in food insecurity, health hazards and forced migration. The marginalized and vulnerable population will mostly bore the brunt of these disproportionately.

Bangladesh is a climate-vulnerable country where the prevalence of poverty is high. The situation has been further exacerbated by the COVID-19 pandemic, which

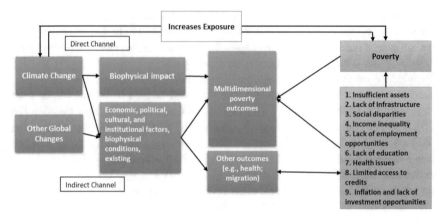

Fig. 3 Complex interactions of climate change and poverty

is now more than a health crisis, as it will have a long-lasting devastating impact on social, economic and political [36] condition of the country, and the recurring natural disasters like cyclone, flood, which are frequent to this region, is further worsening the situation. On the other hand, nationwide lockdown imposed by the governments of the countries to reduce the number of cases of COVID-19 incidence has resulted in global economic shrinkage [2, 44]. Looming global recessionary pressure will have a crippling effect on the lives and livelihoods of the people and will exacerbate if the pandemic continues for a long time [37]. The economic crisis thereby is likely to trigger financial crises for both corporate and household levels [7]. Poverty, hunger, mass starvation, malnutrition, inequalities, gender-based violence, discrimination against poor, marginalized and vulnerable groups are likely to increase due to plummeting socio-economic crises. Moreover, climate-related disasters will add more burden to the already distressing situation. Although, the experience of the country in dealing with natural disasters is appreciated around the world, the compounding effect of both pandemic and disasters will surpass the capacity of the country which in the long-term will affect the achievement of SDG [37]. Also, lack of social and political inclusions followed by climate vulnerability and disasters in Bangladesh may result in the country to fall back into poverty level, and the country can have a hard time recovering from this shock [20].

4 COVID-19 Pandemic-Induced Poverty

Since the general lockdown (which started on 26 March 2020) in Bangladesh, severe constriction of economic activity has pushed more and more Bangladeshis down the poverty line [38]. An analysis by the South Asian Network on Economic Modeling (SANEM) identified that the poverty rate of Bangladesh may have doubled to 40.9% from the onset of the pandemic [38]. To add more to it, COVID-19 has been severely affecting the income of the vulnerable, as people at the bottom of the economic pyramid are facing the severe brunt of job loss. Bangladesh Economic Association (BEA) identified that about 36 million people have lost their job and a huge change took place in class structures in terms of socio-economic conditions. Nearly 59.5 million people moved into different class structures during the lockdown period, of which 25.5 million people became extremely poor [39]. Another study by the Bangladesh Institute of Development Studies (BIDS) shows that by the end of 2020, the country will have 16.4 million new poor due to a reduction in the income of the working-class population in both urban and rural areas. BRAC via a sample survey in May 2020 found that 51% of households had their income reduced to almost zero, and 26% of households have no idea on how to cope with the situation. It has been found that the per capita income of the people living in the slums has dropped by 82%. In February, when the country was fully functional, the earning of an individual was about 108 BDT ($1.30) which decreased to 27 BDT ($0.32) by May 2020 [14]. The slum dwellers of the country were already in a dire situation, which became worse over the pandemic, exasperating the livelihood of these people. The "new poor" in urban areas who have lost their source of income is now returning to their hometowns or rural areas. This kind of reverse migration will overburden the rural economy [40] until the pandemic recedes from the country. Furthermore, the rural economy of Bangladesh is anticipated to be more vulnerable and pressured when 600,000 expatriates who are mostly from rural areas will not migrate back [40]. Therefore, loss of job, decrease in income and lack of physical access to basic public services will have direct implications on people's food security, water and sanitation issues, social protection, health and education system [41].

The change in the poverty pattern is due to the nationwide lockdown imposed by the governments of the countries to reduce the number of cases of coronavirus thereby resulted in global economic shrinkage as well as destabilization of the economy, as the unemployment rate in both formal and informal sector rose greatly with closing down of several small business enterprises (SME), decreasing foreign income from remittance, agricultural sector, readymade garments (RMG) industries, tourism, exports, etc. [38, 42]. The agricultural sector of the country is the single largest economic sector, accounting for 18.6% of the country's GDP, yet it has undergone a severe loss during the 45 days of the lockdown period in the year 2020, likely to have serious implications on the major macro economy of the country [43]. A study by BRAC revealed that between March to May the farmers in Bangladesh had faced a loss of 565.36 billion BDT ($6.66 billion) [38, 44], which in turn will affect the national food security of the country. Also, about 7.2 million agricultural labourers, who are seasonal migrants and usually travel to other parts of the country for alternative livelihood options are, however, got trapped and unemployed due to the pandemic [45]. Physical Capital and assets of households play an important role in analysing its economic and physical vulnerability during the crisis. Moreover, to add to the burden, the prices of essential commodities spiked during the lockdown due to disruption of the supply chain and markets. This, thereby, decreases the people's ability to purchase food as 85% of the country's population earn less than USD 6 (dollar) per day according to World Bank [46], increasing the possibility of a high incidence of malnutrition in near future [47, 48]. According to [45] about 35% of the population belongs to the 5001–10,000 BDT income group and the average household income of eight-division of Bangladesh during January and February before the lockdown was an average of 12,704 BDT and their monthly household expenditure was an average of 13,432.00 BDT. Therefore, it is quite apparent that these households made no savings before the lockdown, thereby is undergoing an acute monetary crisis during this pandemic.

Besides the lower and middle-income group, COVID-19 has severely affected businesses, and organizations throughout the country. An assessment by [45] shows that 55.88% of City Corporations, 39.89% Pourashavas and 38.49% of rural areas businesses and organizations have reported being fully affected. Also, limited movement of transportation around the globe and the nationwide lockdown has dilapidated the global supply chain which in turn has a serious implication on import and export sectors. The RMG sector of Bangladesh, which makes it a major source of income and foreign earnings and contributes significantly to employment generation has also been ramified by the pandemic [49]. This sector alone contributes 85% of the total export of the country and previously for the fiscal year of 2013–2014, the RMG sector accounted for 14.07% share of the country's GDP and 81% of total export earnings [50]. It is well known that the growth of this sector will transcend the economic development of the country while reducing poverty. However, the RMG sector underwent a major strain since the COVID-19 started as international retailers (of UK and USA) started claiming bankruptcy and is undergoing pay cuts, they closed down many of their outlets and apparel factories in Bangladesh [51]. Continuous cancellation of the orders and deals left the factories with half of their normal day orders. Bangladesh Garments Manufacturing and Exporters Association (BGMEA) has estimated that they had lost over 3.7 billion USD since the pandemic started due to order cancellations and refusal of shipments by the USA and UK [51]. In addition, more than 100,000 RMG workers in Bangladesh have been left jobless [52]. Informal sectors also suffered the brunt of the pandemic. The sector, which accounts for 80–90% of the jobs of the country, has been so devastated by the pandemic that their income subsided by as much as 74% [38]. Therefore, the income of large number of populations has decreased significantly with the increase in the number of unemployed populations [48]. Previous record of 10% poverty will go up by 40% according to an analysis by the Centre for Policy Dialogue (CPD) [53]. As the poverty rate doubled and poor people are becoming poorer, this will push back the poverty state of the country to the situation that was faced by the country 15 years ago; setting back the progress of SDGs until now. If such trends or unavailability of work and unemployment trends persists, Bangladesh is going to face greater economic fallout for a long time [13], which will increase the inequalities further.

5 Strategies and Policies for Poverty Reduction in Bangladesh

The government has adopted several medium to long-term strategies to ensure economic progress, address climate change issues and challenges due to the pandemic (shown in Fig. 4).

Poverty is multidimensional in nature thereby it encompasses various deprivations experienced by poor people. Bangladesh government aims to address poverty in all dimensions ensuring that no one is left behind. Therefore, it needs a comprehensive programme for the reduction of poverty and ensures rapid, resilient growth which is both inclusive and sustainable. The 8th five-year plan of Bangladesh has been designed in such way to achieve the vision as well as the sustainable development goals. General Economics Division [18] shows that accelerated GDP growth rates, sectoral transformation from agricultural to non-agricultural sector, creation of employment opportunities, increase of external employment and flow of remittance along with access to microcredits, social protection schemes, etc. has played an important role in poverty reduction. In 2015, National Social Security Strategy (NSSS) has been undertaken to support poverty reduction through a wide range of social safety nets. Some other initiatives like Amar Bari Amar Khamar, Ashrayan Project, Digital Bangladesh, Education Assistance Trust, Women Empowerment Programme, Electricity for All, Community Clinics and Mental Health Programme, Social Safety Net Programme and Investment Development has successfully contributed to the alleviation of poverty. Another notable initiative of government was Strengthening Women's Ability for Productive New Opportunities (SWAPNO) which focuses on poor to extremely poor women of rural areas which help them to fight poverty in an efficient way. The government also made huge investments in human development, reduce gender disparity, ensure financial inclusion and macroeconomic stability. There was also an "Environment Protection" initiative, which addresses the adverse risks of global climate change and aims to the conservation of the environment and biodiversity through sustainable development.

Since climate change affects poor people disproportionately and sets back the progress in poverty reduction, it has always been a priority of the government to reduce the adverse impacts and vulnerability of the population towards climate change. The government is on its way to achieve the targets of SDGs in reducing the deaths and economic losses from natural disasters. They are also developing plans, policies and strategies to promote climate change adaptation and mitigation in order to adapt to the increasing impact of climate change in future [18, 54]. In light of it and also to support international domains effort to address climate change, the government also prepared and submitted a National Adaptation Plan of Action (NAPA) in 2005 to UNFCCC to prioritize the immediate and urgent needs, which is going to be updated to address medium and long-term adaptation planning and programming through National Adaptation Plans (NAP) in 2019. In 2009, Bangladesh Climate Change Strategy and Action Plan (BCCSAP), which is more comprehensive, has been adopted followed by the formation of the Bangladesh

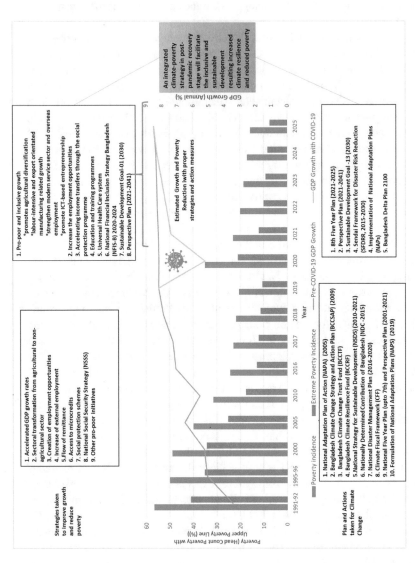

Fig. 4 The figure shows observed and estimated GDP growth before and after the COVID-19 pandemic along with changes in poverty and extreme poverty incidence. Climate change and poverty reduction action measures taken over the time is also included along with possible response mechanism in post-pandemic recovery stage in order to eradicate poverty through sustainable measures.

Climate Change Trust Fund (BCCTF). The government aims to reduce GHG emission by 5% from business-as-usual (BAU) levels by 2030, therefore in 2015, the Nationally Determined Contribution of Bangladesh (NDC-2015) under the Paris Agreement (PA) has been formulated which includes both unconditional and conditional emissions reduction goals for a different sector. Also, the government adopted Sendai Framework for Disaster Risk Reduction (SFDRR) into several policies and introduced a National Plan for Disaster Management in 2016 as strategic guidance for stakeholders to facilitate the understanding of the risks and implementation of the priority actions. Besides, climate change adaptation is mainstreaming into planning processes such as the five-year national development plan (6FYP, 7FYP, 8FYP), National Strategy for Sustainable Development (NSDS) (2010–2021), Climate Fiscal Framework (CFF) which supports climate change-related actions, integrate adaptation planning and identify the demand and supply sides of climate fiscal funds. They have also increased their commitment and investment towards green growth and green technologies, emphasizes the blue economy, implementation of Montreal Protocol on Ozone Depleting Substances (ODSs), development of a pro-poor Climate Change Management strategy, etc. Moreover, the Bangladesh government took a huge step towards climate change resilience through a holistic and integrated Delta Plan 2100, which includes major policies, investment programmes and institutional reforms to reduce the vulnerability to natural hazards and climate change. It is the first time, climate change issue has been considered in developing the macroeconomic framework of the plan, which if implemented will help us to achieve national-level goals of eradicating extreme poverty by 2027 (if the Delta Plan policy option is adopted) and 2041 (in case of Business As Usual policy option) [55]. Other actions that will be taken within the 8FYP are mobilization and utilization of resources for climate funds, governance of climate finance, formulation and advancement of NAP processes, technology transfer on adaptation and mitigation, gender-inclusive Climate Change Response Framework, and increase of public–private partnership.

However, COVID-19 has a devastating impact on the progress of poverty reduction. It had made poor people poorer and drag extreme poor into destitute. The government had taken some notable response mechanisms. Bangladesh has long-standing experience in preparing, facing and overcoming disasters. Therefore, the Ministry of Disaster Management and Relief (MoDMR) took protocols following the Standing order on Disaster (SOD) to address this multifaceted crisis. Corona Prevention committee and local disaster response team were activated at Upazila level under districts and Upazila administration [45]. The government has also introduced an assistance package to minimize the impact of COVID-19 on different sectors and revive the economy of the country from stagnation. The package involved giving RMG sectors' exporting firms, a wage bill support of 50.0 billion BDT in credit at an interest rate of 2.0% implying a subsidy of 7.0% from the market rate of 9.0% (scheduled for operation from April 1, 2020) [49]. Later, this economic stimulus was further promoted to four packages totalling 677.50 billion BDT, which were directed for large businesses, small business enterprises, export development fund and import financing schemes at a low interest rate. Moreover, GoB allocated a stimulus package of 50 billion BDT for agro loans to ensure that rural areas can engage more people in economic activities [49]. Besides the government, various NGOs and communities in Bangladesh also came forward to support the vulnerable communities. Organizations such as BRAC, World Vision, Friendship Bangladesh, Oxfam Bangladesh, Caritas Bangladesh, Jago Nari, Coast, etc. have actively worked for WASH, capacity and awareness building. BRAC, due to its experiences working in a humanitarian crisis has operated effectively and efficiently all over Bangladesh raising awareness on protective wear, hygiene products, and distributing cash relief of a total of BDT 15 million [45]. Development partners like USAID have provided nearly $37 million to support COVID-19 response efforts to assist low-income urban residents of slums in Dhaka city with emergency health, humanitarian, economic, and development assistance [56]. They also provide logistic support and made an investment in PPE productions and other medical supplies essential for the treatment of COVID-19 [57]. A 2-day training course for 1000 Bangladeshi doctors has been carried out by a joint venture of USAID and Directorate General of Health Services (DGHS) to capacitate them to treat infected persons effectively while keeping themselves safe and helping to reduce the spread of the virus. Assistance has also been received from World Bank (WB), the UK government, Asian Development Bank (ADB), IMF and SAARC Development Fund (SDF) to strengthen the national systems for public health emergencies and respond to the COVID-19 (Coronavirus) pandemic [58]. Besides, the UN system is working on addressing critical social and economic impacts, strengthening crisis management through a comprehensive, equitable and inclusive policy response and supporting the health response to eradicate poverty, reduce inequalities and build resilience to present and long term crises and shocks along with the continuation of progress in achieving the SDGs [36]. UN along with other partners is planning to support the country at both national and sub-national levels to address the social concerns and ensure an effective and efficient response to preserve stability and tackle the pandemic.

Now, in the post-pandemic recovery stage, the government is planning to strengthen the robustness of the poverty reduction strategy through pro-poor and inclusive growth through the 8th five-year plan [18] that promotes agricultural diversification, labour-intensive and export-orientated manufacturing-related growth, strengthen modern service sector and overseas employment, promote ICT-based entrepreneurship, etc. The government will also take actions to increase the employment opportunities, ensure resilience and stability of income by increasing the real incomes and accelerating income transfers through the social protection programme which is assumed to reverse the COVID-19 induced poverty in such a way that there will be a balance between regional development. In order to ensure financial inclusion, gender parity and reduced inequality, National Financial Inclusion Strategy Bangladesh (NFIS-B) has been undertaken. The 8th five-year plan also aims to achieve the targets of SDG-1 that is eliminating extreme poverty by 2031. Therefore, it has provision for education and training programmes as well as means of social protection for the extreme poor. Moreover, the government will undertake prevention and mitigation strategies to address shocks like the COVID-19 pandemic, climate change and natural disaster-related shocks to prevent the population to further slip into poverty. Therefore, the government plans to minimize the effects of anticipated shocks through the implementation of Delta Plan 2100 and the introduction of a Universal Health Care system to ensure health safety.

6 Reducing Poverty Through Enhancing Resilience to Climate Change and Post-Pandemic Recovery

COVID-19 reveals the underlying weaknesses in our system and it shows that we are not resilient enough to fight against the pandemic. As the pandemic is derailing the progress of poverty reduction, by creating a pool of "new poor" and making poor people poorer and leading extreme poor into destitute, it shows that the poverty reduction strategies undertaken till date are not sustainable enough to absorb the shocks like COVID-19 pandemic. The population is unable to bounce back from the crisis, requiring the government to provide relief, social safety nets and stimulus packages to support the population and revive the economy. This, however, makes us think about the poverty outcome if this pandemic continues or a new unprecedented shock emerges from a natural disaster or climate change-induced extreme events. Therefore, to make the economy robust and resilient, it is important to ensure that all sectors of the economy contribute to poverty reduction. Hence, to address this crisis within crises, climate action and economic growth needs to be complementary in case of planning and implementation of the long-term development plans.

The pandemic has provided an opportunity to reorient the economies towards a low-carbon trajectory simultaneously addressing the underlying vulnerabilities and improving the resilience of the communities towards climate change and natural disasters. Therefore, the post-COVID recovery plans for the revival of the economy of the country should include the scopes for integration of climate change strategies that support mitigation and adaptation actions. It should also involve massive fiscal stimulus to address COVID-19 impacts, which will provide an opportunity to fast-track sustainable development if they are aligned with the SDGs since it will act as a valuable tool with a comprehensive framework to address risk factors holistically and build resilience [17]. Several climate policy-makers are advocating post-pandemic recovery plans be included in the policies, which will avoid impeding the progress on climate change and serve the dual purpose of ensuring the growth and decarbonizing the economy simultaneously by emphasizing both economic multiplier and climate impact metrics [59, 60]. The government should move back to "build back better" by increasing investments in clean and green physical infrastructure, building efficient retrofits, investing in areas of education and training, natural capital for enhancing ecosystem resilience, micro-financing and research and development with poverty reduction at the centre of the policy response. Moreover, a coordinated and integrated approach should be adopted which will ensure that all response measures are inclusive, efficient, gender-sensitive, alleviate poverty and address inequality, have business continuity measures, strengthen public service delivery, increase resilience and help to attain sustainable development goals (SDGs) [1]. Also, a strong and effective framework should be built to ensure basic rights, adequate access to public health infrastructure, provision of clean water and sanitation, increased investments in building digital capacity and ensuring availability of digital platforms for education and financial services. Furthermore, sustainable development should be ensured through public and private partnership and investment in building resilient infrastructure; strengthening social protection systems; planning effective risk reduction and crisis prevention strategies and removing any barriers that affect supply chains.

7 Conclusion

Poverty is a dynamic and multidimensional condition, which has been further worsened in Bangladesh due to crises like the COVID-19 pandemic and climate change vulnerabilities. Before the current pandemic, Bangladesh was falling behind the achievement of targets of sustainable development goals. Now, COVID-19 is threatening to derail the progress of the country in eradicating poverty (SDG-1). The country has also failed to put enough efforts to prevent the transmission of the virus due to late apprehension of the risk, poor planning, weak governance structure, regulatory framework, centralized healthcare system, insufficient institutional capacity and management system, absence of timely and effective decision-making, fragmented public services delivery and inefficient allocation of public resources. This has threatened the poor population and those just above the poverty line to slip into extreme poverty creating another group of "new poor". It also threatens the return of expatriates, a decrease of remittance, a decrease of employment opportunities, which will cause reduction of income of the population, dragging them down the poverty line. Moreover, the climate change crisis is further intensifying poverty by increasing people's exposure and vulnerabilities towards natural and climate change-related disasters. This pandemic provides a great opportunity to include climate change-related issues into post-pandemic development planning which aims to eradicate poverty and achievement of sustainable development goals by adopting a pro-poor and inclusive poverty reduction strategy along with the promotion of low-carbon and green economic growth.

In order to address these complex and multidimensional crises, it is important to prioritize the areas and formulate a set of policies and solutions to respond effectively and efficiently towards a sustainable future. Therefore, all dimensions of poverty such as lack of income, consumption, lack of access to health and education and vulnerability towards any shocks or stresses should be addressed through climate-integrated poverty reduction efforts. To ensure adequate implementation of the post-pandemic recovery plan and to address the new dimension of poverty, it is important to ensure inter-sectoral coordination and collaboration, public–private partnership, access to safety net programmes, development of a comprehensive plan and holistic guidelines, guaranteeing transparency and accountability. This will reduce the future risk towards climate change and other unprecedented crises like the COVID-19 pandemic by making the community and population resilient and self-sustaining through eradication of poverty and undertaking sustainable strategies.

References

1. United Nations (2020a) B R I E F # 2: putting the UN framework for socio-economic response to COVID-19 into action: insights (Issue June).
2. Karim MA (2020) SDG implementation for post-pandemic economic recovery. The Business Standard. https://www.tbsnews.net/analysis/sdg-implementation-post-pandemic-economic-recovery-101212?fbclid=IwAR2Dk_uheUUWq5TkYPU8AZuyXuiw-WkFB02h4vTSZLYu JjYvO5vZkJV31yA
3. Solberg E, Akufo-Addo NAD (2020) How is COVID-19 affecting Sustainable Development Goals around the world? World Economic Forum. https://www.weforum.org/agenda/2020/04/coronavirus-pandemic-effect-sdg-un-progress/
4. Mukarram M (2020) Fighting pandemic with SDGs in mind. The Financial Express. https://thefinancialexpress.com.bd/views/views/fighting-pandemic-with-sdgs-in-mind-1591803501
5. Dhakal M (2020) COVID-19 another shock for vulnerable countries facing multiple crises. Cimate Analytics . https://climateanalytics.org/blog/2020/covid-19-another-shock-for-vulner able-countries-facing-multiple-crises/
6. Roy P, Hasan R, Alamgir M (2020) Amphan inflicts massive damage. The Daily Star. https://www.thedailystar.net/frontpage/news/amphan-inflicts-massive-damage-1904977
7. Ramachandran S (2020) The COVID-19 Catastrophe in Bangladesh. The Diplomat. https://thediplomat.com/2020/04/the-covid-19-catastrophe-in-bangladesh/
8. Huq S, Rabbani G (2011) Climate change and Bangladesh: policy and institutional development to reduce vulnerability. J Bangladesh Stud 13(September):1–10
9. Eckstein D, Künzel V, Schäfer L, Winges M (2019) Global climate risk index 2020. Germanwatch e.V., Kaiserstr
10. Mahmud I (2019) Bangladesh ranks 3rd among countries most hit by disaster, 7th in climate vulnerability. Prothom Alo. https://en.prothomalo.com/bangladesh/Bangladesh-ranks-3rd-among-countries-most-hit-by
11. ADB (2021) Poverty: Bangladesh. Asian Development Bank (ADB). https://www.adb.org/cou ntries/bangladesh/poverty
12. Anner M (2020) Abandoned? The impact of Covid-19 on workers and businesses at the bottom of global garment supply chains executive summary. https://www.ilo.org/global/about-the-ilo/newsroom/news/WCMS_738742/lang--en/index.htm
13. Bhuiyan AKMI, Sakib N, Pakpour AH, Griffiths MD, Mamun MA (2020) COVID-19-related suicides in bangladesh due to lockdown and economic factors: case study evidence from media reports. Int J Ment Heal Addict. https://doi.org/10.1007/s11469-020-00307-y
14. Kamruzzaman M (2020) Coronavirus: Poor income drops 80% in Bangladesh. Anadolu Agency. https://www.aa.com.tr/en/asia-pacific/coronavirus-poor-income-drops-80-in-bangla desh/1808837
15. Huq S (2020) Covid-19 and climate change. The Daily Star. https://www.thedailystar.net/opi nion/politics-climate-change/news/covid-19-and-climate-change-1885357
16. Hasina S, Verkooijen P (2020) Fighting cyclones and coronavirus: how we evacuated millions during a pandemic I Global development. The Guardian. https://www.theguardian.com/global-development/2020/jun/03/fighting-cyclones-and-coronavirus-how-we-evacuated-millions-during-a-pandemic?fbclid=IwAR3iIQnXDIDl3keyaabcCC5JlJkLsMGWnfEVLkyczsA_Fzo hgSSN4lCrca8
17. Moores D (2020) With COVID-19, the SDGs are even more important: Devpolicy Blog from the Development Policy Centre. Development Policy Centre (DevPolicyBlog). https://devpol icy.org/with-covid-19-the-sdgs-are-even-more-important-20200616-2/

18. General Economics Division (GED) (2020a) 8th Five Year Plan (July 2020-June 2025): Promoting Prosperity and Fostering Inclusiveness. IRC
19. World Bank (2020) Globally Bangladesh is a model for poverty reduction. The World Bank. https://www.worldbank.org/en/news/press-release/2020/01/29/globally-bangladesh-is-a-model-for-poverty-reduction-world-bank
20. Citizen's Platform for SDGs-Bangladesh (2019) Four years of SDGs in Bangladesh and the way forward. https://cpd.org.bd/wp-content/uploads/2019/07/Four-Years-of-SDGs-in-Bangladesh.pdf
21. General Economics Division (GED) (2019) Banglaesh moving ahead with SDGs. Prepared for Bangladesh Delegation to 74th UNGA Session 2019
22. General Economics Division (GED) (2020c) Sustainable development goals Bangladesh progress report 2020.
23. Government of the People's Republic of Bangladesh (2020) Voluntary National Reviews (VNRs) 2020, accelerated accelerated action and transformative pathways: realizing the decade of action and delivery for sustainable development (Issue June).
24. UNDESA (2018) Leaving the LDCs category: Booming Bangladesh prepares to graduate. United Nations Department of Economic and Social Affairs (UNDESA). https://www.un.org/development/desa/en/news/policy/leaving-the-ldcs-category-booming-bangladesh-prepares-to-graduate.html
25. Byron RK, Mirdha RU (2021) Becoming a developing nation: Bangladesh reaches a milestone. The Daily Star. https://www.thedailystar.net/frontpage/news/becoming-developing-nation-bangladesh-reaches-milestone-2052161
26. IISD (2018) Bangladesh, UN consider expected LDC graduation in 2024 I News I SDG Knowledge Hub. International Institute for Sustainable Development. https://sdg.iisd.org/news/bangladesh-un-consider-expected-ldc-graduation-in-2024/
27. General Economics Division (GED) (2020b) Perspective plan of Bangladesh 2021–2041: making vision 2041 a reality (Issue March). http://www.plancomm.gov.bd/sites/default/files/files/plancomm.portal.gov.bd/files/10509d1f_aa05_4f93_9215_f81fcd233167/2020-08-31-16-08-8f1650eb12f9c273466583c165a315a4.pdf
28. Charles A, Kalikoski D, Macnaughton A (2019) Addressing the climate change and poverty nexus: a coordinated approach in the context of the 2030 agenda and the Paris agreement. FAO, Rome. https://doi.org/10.1108/acmm.2001.12848daf.003
29. Leichenko R, Silva JA (2014) Climate change and poverty: vulnerability, impacts, and alleviation strategies. Wiley Interdiscip Rev Climate Change 5(4):539–556. https://doi.org/10.1002/wcc.287
30. Eriksen SH, O'Brien K (2007) Vulnerability, poverty and the need for sustainable adaptation measures. Climate Policy 7(4):337–352. https://doi.org/10.1080/14693062.2007.9685660
31. IPCC (2012) Special report of the intergovernmental panel on climate change managing the risks of extreme events and disasters to advance climate change adaptation.
32. Kelly PM, Adger WN (2000) Theory and practice in assessing vulnerability to climate change and facilitating adaptation. Clim Change 47(4):325–352. https://doi.org/10.1023/A:1005627828199
33. Rayner S, Malone EL, Rayner S, Malone EL (2001) Climate change, poverty, and intragenerational equity: the national level. Int J Global Environ Issues 1(2):175–202. https://doi.org/10.1504/IJGENVI.2001.000977
34. Kim N (2012) How much more exposed are the poor to natural disasters? Global regional measurement. Disasters 36(2):195–211. https://doi.org/10.1111/j.1467-7717.2011.01258.x
35. Roy J, Tschakert P, Waisman H, Halim SA, Antwi-Agyei P, Dasgupta P, Hayward B, Kanninen M, Liverman D, Okereke C, Pinho P, Riahi K, Suarez Rodriguez A (2018) Sustainable development, poverty eradication and reducing inequalities. Special Report, Intergovernmental Panel on Climate Change, pp 445–538. https://www.ipcc.ch/site/assets/uploads/sites/2/2019/05/SR15_Approval_Chapter_5.pdf
36. UNDP (2020) Coronavirus disease COVID-19 pandemic I UNDP in Bangladesh. United Nations Development Programme (UNDP). https://www.bd.undp.org/content/bangladesh/en/home/coronavirus.html

37. Mahmood M (2020) Covid-19: economic challenges facing Bangladesh. The Financial Express. https://thefinancialexpress.com.bd/views/covid-19-economic-challenges-facing-ban gladesh-1592064588
38. Ahmed S (2020) Covid-19 pandemic: implications for Bangladesh. The Business Standard. https://tbsnews.net/thoughts/covid-19-pandemic-implications-bangladesh-82285
39. Bhuiyan MSA (2020) Covid-19 and its impact on Bangladesh economy. The Business Standard. https://tbsnews.net/thoughts/covid-19-and-its-impact-bangladesh-economy-69541
40. UNB (2020) Experts: Rural economy going to be under pressure with returnees from cities. Dhaka Tribune. https://www.dhakatribune.com/business/economy/2020/07/11/experts-rural-economy-going-to-be-under-pressure-with-returnees-from-cities
41. HCTT (2020) Humanitarian preparedness and response plan for climate-related disasters in 2020 (Vol. 2020, Issue May).
42. Rahman M (2020) Bangladesh: Navigating the globalized impact of the pandemic. Friedrich Ebert Stiftung. https://www.fes-asia.org/news/bangladesh-navigating-the-globalized-impact-of-the-pandemic/
43. Islam J, Babu MU (2020) Coronavirus in Bangladesh: Bangladesh rural economy reels from shutdown. The Business Standard. https://tbsnews.net/economy/rural-economy-reels-shu tdown-63994
44. Ahmed Z (2020) Coronavirus: economy down, poverty up in Bangladesh I Asia I An in-depth look at news from across the continent I DW I 10.06.2020. Deutsche Welle (DW). https://www.dw.com/en/coronavirus-economy-down-poverty-up-in-bangladesh/a-53759686
45. NAWG (2020) Needs Assessment Working Group (NAWG) Bangladesh I HumanitarianResponse. UNOCHA. https://www.humanitarianresponse.info/en/operations/bangladesh/needs-assessment-working-group
46. Saleh A (2020) Why Bangladesh is especially vulnerable to the coronavirus. World Economic Forum. https://www.weforum.org/agenda/2020/04/in-bangladesh-covid-19-could-cause-a-humanitarian-crisis
47. Hasan MR (2020) Amid Covid-19 Hunger Fear Mounts in Bangladesh. Inter Press Service (IPS). http://www.ipsnews.net/2020/04/amid-covid-19-hunger-fear-mounts-bangladesh/
48. Uttom S (2020) Covid-19 fuels hunger and poverty in Bangladesh. UCA News. https://www.ucanews.com/news/covid-19-fuels-hunger-and-poverty-in-bangladesh/88343
49. Taslim MA (2020) Covid-19 pandemic and government response. The Financial Express. https://thefinancialexpress.com.bd/views/views/covid-19-pandemic-and-government-res ponse-1588000220
50. Islam MS, Md. Abdur Rakib, Adnan A (2016) Ready-made garments sector of bangladesh: its contribution and challenges towards development. J Asian Dev Sci 5(2):17–26
51. Uddin J (2020) Running at half steam. The Business Standard. https://www.tbsnews.net/eco nomy/rmg/be-blip-98536
52. BIGD (2020) Covid-19 Impact on RMG Sector and the Financial Stimulus Package: Trade Union Responses (Vol. 01, Issue February 2019). https://bigd.bracu.ac.bd/wp-content/uploads/2020/05/Final_Merged_summary_Bangla-English_TU_with-all-feedback.pdf
53. CPD (2020) An analysis of the national budget for FY2020–21. CPD Budget Dialogue 2019, Issue June
54. Huq S, Khan MR, Brief P (2017) Planning for adaptation in Bangladesh: past, present and future, Issue August, pp 1–4
55. Alam S (2019). Bangladesh delta plan 2100: Implementation challenges and way forward. The Financial Express. https://thefinancialexpress.com.bd/views/reviews/bangladesh-delta-plan-2100-implementation-challenges-and-way-forward-1553354695
56. USAID (2020b) United States government provides $7 million in food assistance to help urban poor in Dhaka Cope with COVID-19 challenges I press release I Bangladesh I U.S. Agency for International Development. USAID Bangladesh. https://www.usaid.gov/bangladesh/press-rel eases/jul-9-2020-united-states-government-provides-7-million-food-assistance
57. USAID (2020a) The United States Government delivers more than $173 million in new funding to support Bangladesh's COVID-19 response efforts and post-COVID development

and economic recovery I press release I Bangladesh I U.S. Agency for International Development. USAID Bangladesh. https://www.usaid.gov/bangladesh/press-releases/jun-15-2020-united-states-government-delivers-173-million-new-funding-covid-19-response

58. KPMG (2020) Bangladesh: KPMG global. KPMG International. https://home.kpmg/xx/en/home/insights/2020/04/bangladesh-government-and-institution-measures-in-response-to-covid.html

59. Hepburn C, O'Callaghan B, Stern N, Stiglitz J, Zenghelis D (2020) Will COVID-19 fiscal recovery packages accelerate or retard progress on climate change? Oxf Rev Econ Policy 36(20):1–48. https://doi.org/10.1093/oxrep/graa015

60. IUCN (2020) IUCN calls on business to put nature at the heart of economic recovery plans. International Union for Conservation of Nature (IUCN). https://www.iucn.org/news/business-and-biodiversity/202006/iucn-calls-business-put-nature-heart-economic-recovery-plans

61. United Nations (2020b) Progress towards the sustainable development goals: report of the secretary-general.

Printed in the United States
by Baker & Taylor Publisher Services